AF737087

THE ART AND SCIENCE OF GREEN ARCHITECTURE

The Art and Science of Green Architecture

WAN LI, LIU XIAOXUE, AND CHI XIN'AN

UNIVERSITY OF TORONTO PRESS
Toronto Buffalo London

© University of Toronto Press 2025
Toronto Buffalo London
utppublishing.com
Printed and bound by CPI Group (UK) Ltd, Croydon, CR0 4YY

ISBN 978-1-4875-6698-2 (cloth)
ISBN 978-1-4875-6700-2 (EPUB)
ISBN 978-1-4875-6699-9 (PDF)

Endless Renewal: The Art and Science of Green Architecture
Written by Wan Li, Liu Xiaoxue, Chi Xin'an
© Originally published by Tsinghua University Press Limited 2023

Library and Archives Canada Cataloguing in Publication

[A catalogue record for this publication is available from Library and Archives Canada.]

Cover design: Design Pickle

We wish to acknowledge the land on which the University of Toronto Press operates. This land is the traditional territory of the Wendat, the Anishnaabeg, the Haudenosaunee, the Métis, and the Mississaugas of the Credit First Nation.

University of Toronto Press acknowledges the financial support of the Government of Canada, the Canada Council for the Arts, and the Ontario Arts Council, an agency of the Government of Ontario, for its publishing activities.

Canada Council for the Arts Conseil des arts du Canada

ONTARIO ARTS COUNCIL
CONSEIL DES ARTS DE L'ONTARIO
an Ontario government agency
un organisme du gouvernement de l'Ontario

Government of Canada gouvernement du Canada | Canada

Preface

Hello, I am Wan Li.

During my studies, although I focused on architecture as my major, I was curious about lots of other fields too, wanting to know a bit about everything. So I often read popular science pieces and sometimes wrote articles, sharing a little of the professional knowledge I picked up. In my work, teaming up with experts from different fields became the norm. From them, I gained tons of interesting insights, looked at my own field from all sorts of angles, and got a clearer sense of my role and place.

Maybe you've felt this: In our time, disciplines split finer and finer, yet the situations we face grow more tangled, needing thoughts from every angle, every level, to come up with solid solutions. We can't master every field deeply and systematically, of course, but getting a broad feel for the knowledge and views across disciplines really helps widen our horizons and spark inspiration. That's why, when the editor reached out, the idea of writing a book like this took root in me.

This book doesn't carry heavy theories or murky language. I want to use plain words and real examples to walk you through the origins, ideas, and practices of green architecture and sustainable development. If you can stretch out a bit of your own thinking from it, or find some joy, I'll be more than satisfied. Because of this starting point, this book might not dive into cutting-edge tech details or list unbuilt latest designs but, instead, kicks off from way back in the past, from why we do this, from those standout cases already built and tested by time. If you're not a professional, I hope this book gives you a full sense of green architecture and sustainable ideas, offering a bit of useful reference when you make choices or take action. If you're a professional, I hope it hands you some fresh thoughts, sparks, echoes, or moments to reflect.

If I saw myself as a teacher, I'd be all nerves right now, scared stiff that what I'm presenting isn't perfect enough. But I'd rather be your guide or buddy, walking alongside you to dig into some intriguing topics we might stumble across on this road. Maybe this trip won't show us every view, maybe we'll hit questions with no answers, but I trust the process will enrich our lives. Plus, more journeys are waiting for you out there.

Are you ready to go?

Wan Li
Summer 2023

Table of Contents

THE ART AND SCIENCE OF GREEN ARCHITECTURE

01 Revelations from Ancient Wisdom

When you hear "green architecture," what's the first picture that pops into your head? A house wrapped in green plants? A house covered with solar panels? Before we dive in, I'd like you to set aside those images in your mind for now, come back with me to the past, and see what big ideas about green architecture humans picked up while living with nature.

01-1 "Carving Cave Dwellings into the Earth" on the Loess Plateau

In Northwest China's Loess Plateau, there's a really old way of living called cave dwellings, and I bet you've all heard of them. Ever wonder how cave dwellings got started? Why did folks here think to live in them instead of treehouses, wooden huts, stone houses, or something else?

Let's first take a look at the areas where cave dwellings are common, and what the natural setup is like there.

The Loess Plateau in the Northwest has dry weather and thick soil layers. How thick? Usually over 50 meters, and the thickest spots hit 300 meters. The local climate's pretty cold in winter and dry and hot in summer, with lots of wind and sand.

People figured out that digging a hole in that thick soil keeps it warm in winter, cool in summer, blocks the wind and sand, and saves a bunch of building materials. So, folks started "carving cave dwellings into the earth" on stable soil layers, and over time, that turned into the cave-dwelling way of living.

Cave dwellings are built with unbaked raw earth, what we call "living soil." Unlike bricks, cement, or steel that need processing and shipping, this material's right there to use, cutting down big time on energy for making and moving it. Even when a cave dwelling's life ends and it's torn down or left behind, it's just dust to dust, earth to earth, and no piles of construction trash or materials that won't break down.

Entrances and courtyards of cave dwellings

Cave dwellings of Lijiashan Village, Shanxi

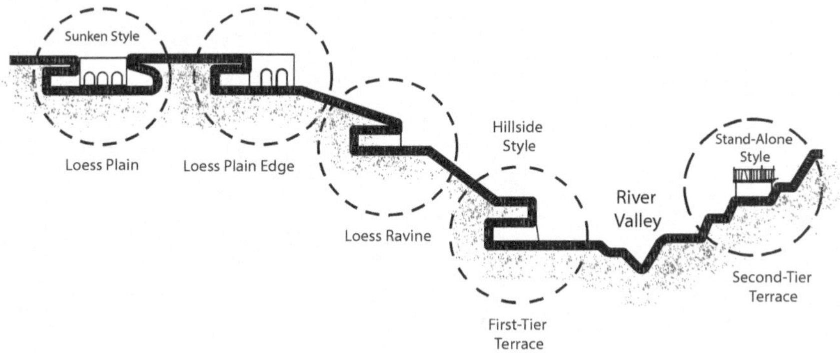

How landforms relate to cave-dwelling types

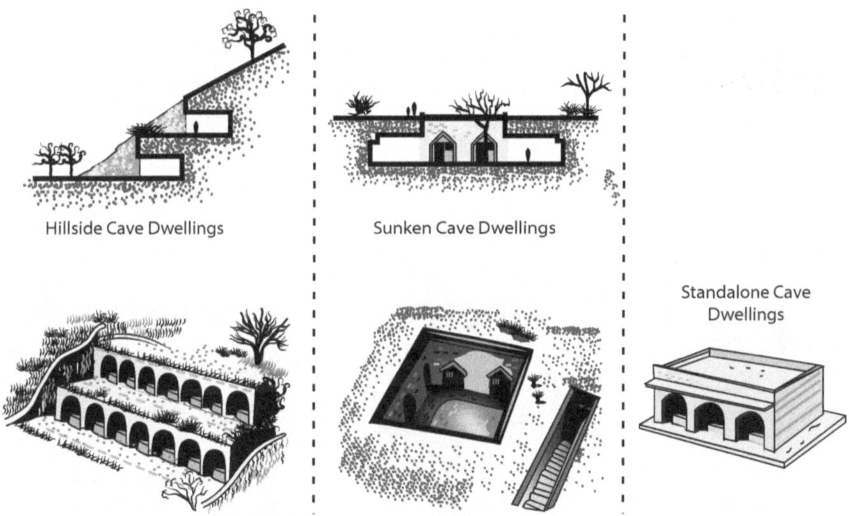

Three types of cave dwellings

Cave dwellings hug the hills and flow with the land. Depending on where they sit, they split into three basic kinds: hillside, sunken, and standalone.

Looking at the local climate, geography, and natural resources, cave dwellings feel like the most "no-brainer" kind of building. Like I said before, inside cave dwellings, it's "warm in winter, cool in summer." How does that warm-in-winter, cool-in-summer thing work? In Shaanbei's cold winters, when it's –20°C outside, inside a cave dwelling it still holds above 10°C. Sunlight, indoor lighting, people moving

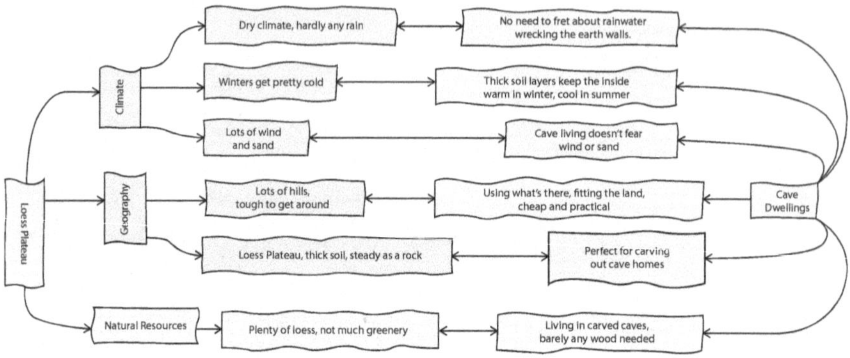

How cave dwellings respond to the local environment

around, cooking in the kitchen, all that gives off heat. Cave dwellings trap that heat inside really well, so in winter you don't need extra heating to keep the place comfy enough.

From a money standpoint, cave-dwelling materials come straight from the ground; there's no fancy processing, and running them doesn't need extra heating, so building and keeping them going costs next to nothing.

From a historical and cultural view, using local materials, local hands, and local know-how, cave dwellings incorporate the unique ways people live and work here, packed with regional flavor.

That's all the one-of-a-kind strengths and charm of cave dwelling.

But as society grows and lifestyles shift, this old system of cave dwellings is facing a slow fade-out crisis. With tech moving forward and living standards climbing, what modern folks want from their homes is changing too. Some cave dwellings have few windows, not enough light inside, poor airflow, and they don't quite fit how people live now. Some people tie cave dwellings to poverty and backwardness and even see tearing them down for brick houses as a sign of getting ahead.

You can see how a building style's edge in one era ties tight to how people lived and what they needed back then. But human society keeps moving, and if building styles and skills don't keep up, innovate, and grow, even popular ones end up on the sidelines.

You Know What?

Ancient folks had no industrial building materials, no mechanized heating or cooling systems. When they built houses, they had to think hard about the local climate, geography, and natural resources.

Building to fit the place, using what's at hand, means working with the local climate, making the most of local resources and energy, and crafting the best living space with what's available.

Sure, the ancients didn't have a worked-out "green architecture" idea, but people living in different spots, using local conditions and quirks, made a wild mix of jaw-dropping local buildings. A lot of the green building tricks we use today were already tucked inside those local structures.

01-2 "One Seal" of the Red Soil Plateau

In Yunnan's red soil plateau in China, the climate's pretty mild, and red earth is everywhere. Folks here often live in houses built from raw earth too. But even though they're earth-built houses, the Southwest and Northwest do it a bit differently.

Yunnan's soil layer isn't as thick as the Loess Plateau, usually just 0.5–2 meters, and red earth is mostly weathered materials, not as steady or as solid as loess. So, while they're still using what's local, people in Yunnan don't carve caves to live in; they use molds and tamping hammers to pack red earth into walls or make adobe bricks to stack up houses.

A common raw-earth home in Yunnan's got a special name, called "One Seal." It's usually square-shaped on the floor plan, with three main rooms, two side rooms on each wing, and a half-outdoor "reverse seat" space at the entrance all boxed together. Normally, a thick outer ring of walls gets built from rammed earth or adobe bricks, while inside, wooden posts, beams, and floors hold it up. The whole courtyard's laid out tight, solid outside, open inside, square like a stamp, so they call it One Seal.

Of course, depending on the land, the people, and the money situation, One Seal rammed-earth courtyard houses come in lots of different twists. But no matter how it varies, the essence stays the same. With the cave-dwelling example in mind, it's easy to see how One Seal connects to the local environment.

Yunnan's One Seal is like the Northwest's cave dwellings, built from what's there, warm in winter, cool in summer, cheap and handy, and full of local character. But just like cave dwellings, it's hitting challenges where space, light, and airflow don't quite cut it for modern living needs anymore. Traditional rammed-earth farmhouses use small gray tiles on the roof that need regular checkups. If checkups lag and leaks spring up, rainwater chews through the walls quickly, doing damage you can't undo. On top of that, Yunnan's a place with lots of earthquakes, and

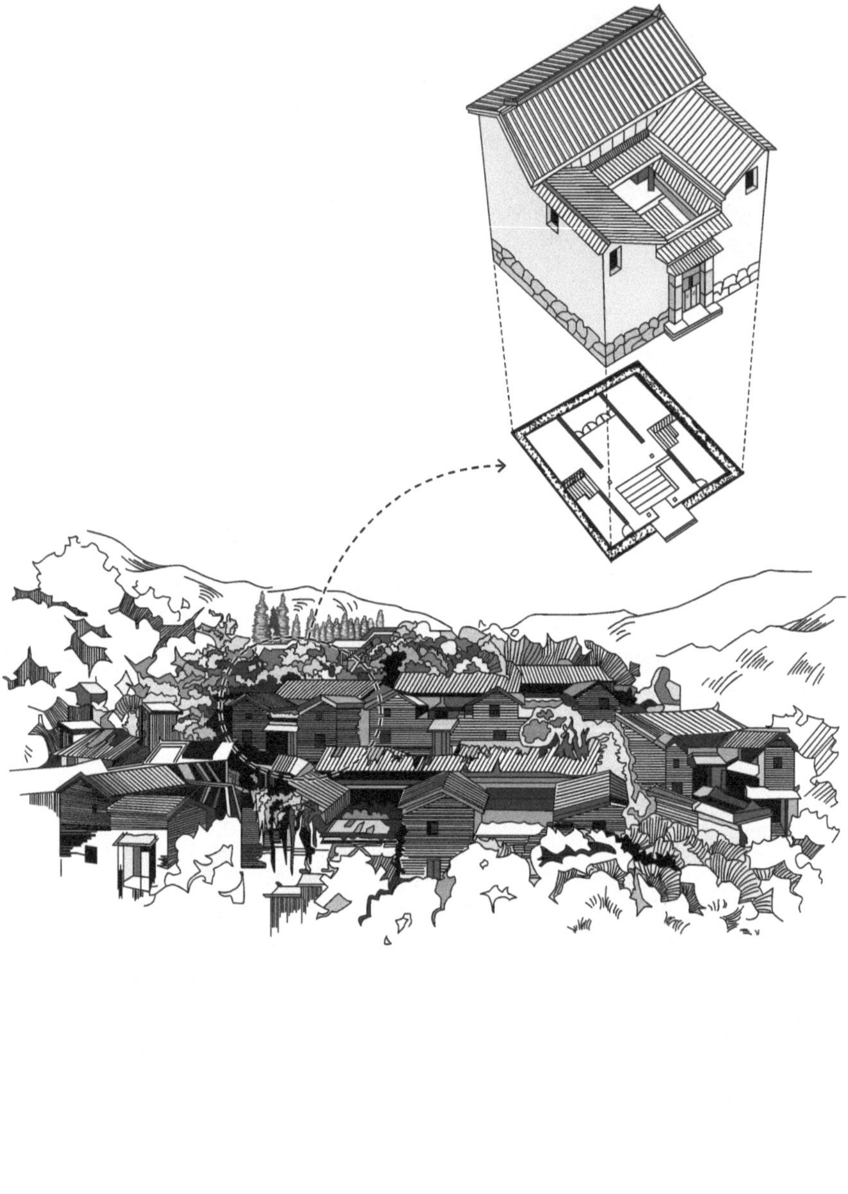

Rammed-earth settlement in Leju Village, Tuanjie Township, Kunming, Yunnan

Yunnan One Seal

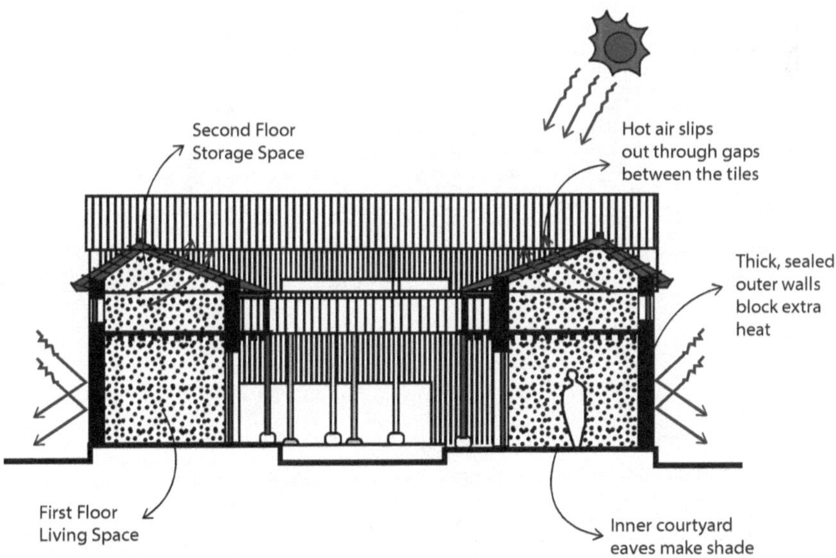

Cross-section diagram of Yunnan's One Seal

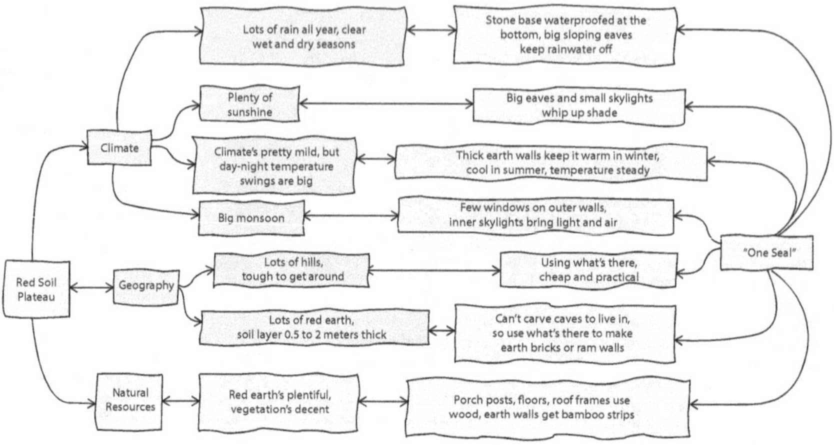

How Yunnan's rammed-earth homes answer to the local environment

old, neglected raw-earth buildings often collapse big-time when they hit, even hurting or killing people. So every time an earthquake shakes things up, a bunch of villagers ditch traditional raw-earth homes and go for brick houses instead. Even with rough roads and skyrocketing costs, for safety's sake, it feels like the only choice local folks have.

Is there really no other way, one that keeps traditional building's good points while fixing comfort and safety? You bet there is! It takes scientific ways to tweak and overhaul those old building strategies.

You Know What?

According to UNESCO stats, at the start of the 21st century, about a third of the world's people still lived in raw-earth buildings. China's 2010–2011 census data show at least 60 million folks in the country lived in raw-earth buildings.

The energy-saving, eco-friendly perks of raw-earth buildings are plain to see:

1. Raw-earth materials don't need firing or heavy processing.
2. Using what's there skips the energy and emissions of long-haul transport.
3. Good insulation and heat-blocking cut energy use and emissions when the building's running.
4. After the building's torn down, the materials break down naturally or get used again.

On top of that, raw-earth buildings carry a rich, colorful slice of local culture. Still, a good chunk of the regions where raw-earth buildings sit happen to be in quake-prone zones, so quake resistance is a big factor holding back their growth and legacy.

01-3 Picking Up the "Baton" from Our Ancestors

Raw earth is a material with low toughness and is decent at handling pressure, but it is weak against bending or pulling. That's why raw-earth buildings without quake-proofing fall apart easily in earthquakes, with roofs, floors, and walls caving in, wrecking property and hurting people. In every quake in Southwest China, spots packed with earth houses take the hardest hits. Statistics show that in the 2014 Yunnan Ludian quake, at least 80% of the collapsed houses were earth-built.

After the August 2014 Yunnan Ludian quake, my team from Hong Kong Chinese University's "One Expert, One Village" joined up with structure and quake experts from Cambridge and Kunming Tech, heading to Guangming Village in Longtoushan Town, Ludian County, Yunnan, to check things out.

No surprise, most of the traditional rammed-earth farmhouses there crumpled in the quake, and villagers started switching to brick houses. But with outside materials and labor costs shooting up after the quake, rebuilding homes turned into a hefty cash load for villagers. Lots of rebuilt houses, short on funds, got no insulation or heat-proofing, no outer wall finishes or indoor fixes—forget furniture, some didn't even have doors or windows put in. Living in those bare redbrick houses, "cold in winter, hot in summer," isn't exactly comfy or decent.

We also found out that, with no know-how on quake-proof building, some of those rebuilt brick houses look sturdy but still have weak spots and safety risks.

Is there a way for villagers to rebuild safe, comfy homes without leaning too hard on outside materials and tech, just using their own resources and skills? We figured that instead of jumping to replace traditional rammed-earth ways, we'd keep an open mind, poke around at what local tech is already there, and see what's worth keeping and what could use a fix.

After digging into the local situation, our expert crew gave traditional rammed-earth buildings a "checkup," spotted the weak points, tweaked them with modern science and tech, and cooked up a new quake-proof rammed-earth farmhouse method that shows some ingenuity, but it's still easy to pick up and use.

Traditional rammed-earth farmhouses collapsed after the Ludian quake

After a bunch of strength tests, our new rammed-earth farmhouse tech proved safe and trusty, meeting all the country's quake-proof building rules. But how do we pass this technique to villagers? Talking endlessly might not do the trick, even if we exhaust ourselves explaining. We decided to "learn by doing" with the villagers, building a demo project so they could feel and grab this technology in the most straightforward way. So we got village craftsmen in on it, designing and building a demo farmhouse for an old couple who couldn't rebuild on their own.

Just fixing safety isn't enough—to shake the villagers' impression that "earth houses mean backward, poor, unsafe," the new farmhouse had to ditch the old rammed-earth problems of cramped space, dim light, stale air, and damp gloom. This means architects must really understand and respect what folks need day-to-day, think hard about fixes, and elevate the entire living situation for villagers.

On a tight little plot, we drew up a snug, single-unit house for the old couple. Two earth-wall-bearing rectangle blocks carry all the daily living rooms for the folks, with a long, narrow mid-court stuck between, lit by a roof hole, a half-outdoor spot that lets daylight and fresh air roll through. It fills in the light and air inside the building very nicely, giving this older couple a spot that keeps out wind and rain while staying bright and cozy for hanging out.

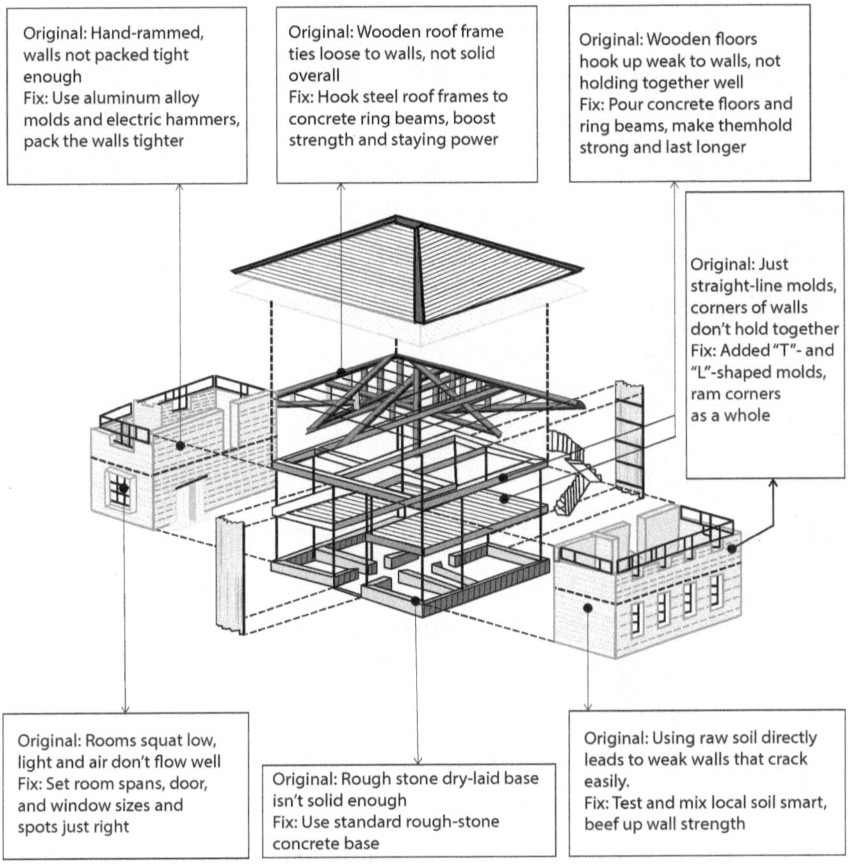

Steps to fix up new rammed-earth farmhouses

Besides safety and comfort, there's another important issue, the economic issue, that needs serious consideration. Before, regular brick-and-concrete houses cost more than the villagers could afford, not to mention most materials came from outside, and skilled workers had to be brought in from far off too. This meant that of the funds invested in materials and labor for construction, only a small portion remained local to support the region's economy. However, the Guangming Village post-earthquake reconstruction demonstration project reversed this approach, utilizing local materials and labor to complete the work and retaining construction funds within the community.

As the new technology enhances local traditional methods with essential improvements, the construction process is straightforward to

Grandma likes doing embroidery in the mid-court

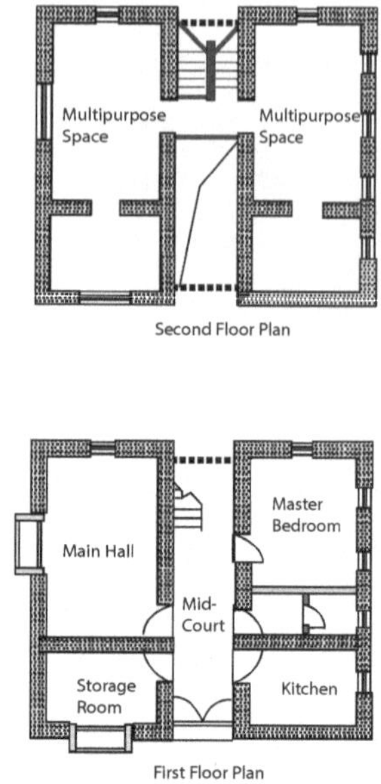

New Home Floor Plan

learn, and the tools are easy to use. With a quick bit of training and hands-on time, a crew of local villagers can take on the house-building job. Villagers, tech guides from the "One Expert, One Village" team, and on-site architects worked and learned together, wrapping up the rebuild in just three months. In addition, the materials needed to build the house mostly came from recycling quake rubble or buying nearby, with all the craftsmen hired local workers. Consequently, construction costs were significantly reduced, to approximately 60% of a conventional brick-and-concrete structure, with the majority of the expenditure remaining local, enhancing the community's income.

Once the project wrapped, the couple moved into their new place. The elderly man gained a well-lit, comfortable space for his skilled bamboo weaving, while his wife, who had not engaged in needlework

Village craftsmen ramming earth walls with the upgraded tools

for years, resumed embroidery with enthusiasm. The post-earthquake reconstruction project not only provided them with a new residence but also restored their confidence and hope for the future, demonstrating that local materials, skills, and labor could construct environmentally friendly, comfortable, cost-effective, and respectable homes.

Huh? Isn't this book about green architecture? Why have I elaborated so much on my team's project, as if I've gone off-topic? In fact, presenting these traditional techniques and local knowledge serves a purpose: Although green architecture often carries a sophisticated stereotype, seeming distant from daily life, advanced technology, new materials, and expensive aesthetics are not its only strengths. Peel back that surface, and green architecture's really about using smart, fitting ways to handle the ties between people and the environment, people and each other and getting those ties into a healthy, lasting loop. This isn't just science and engineering know-how—it pulls in social, human, and economic angles too, calling for real open, bendy thinking and a ton of cross-field, cross-team talk and teamwork.

Guangming Village Post-Quake Rebuild Demo Farmhouse Finished Look

Folks keep asking me, "Why bother with dirt, these backward materials, to build houses?" But I always figure, for buildings in different spots, materials don't pick sides—there's no flat-out good or bad. Every material's got its own quirks and where it fits best. As long as you've got a grip on the local scene and respect for how folks live there, putting different materials where they belong, steel or dirt, can make people's lives and spirits shine fresh. Sometimes, it looks like we're just thinking about the villagers' post-quake rebuild and future livelihoods, but really, we're already living out true green building ideas. The Guangming Village post-quake rebuild demo received the Hong Kong Green Building Award for new homes and a special nod from the United Nations for its Sustainable Development Goals. As a little earth house, the green building ideas packed inside it don't fall short of other award winners one bit.

A German architect I really admire, Anna Heringer, once said: "Architecture is a tool to improve lives." Have you ever thought about the buildings around you from this angle? Next, let's check out those green buildings that "improve lives" together!

02 Dancing with Nature

With the limited tech of the past, our ancestors figured out how to build homes best suited to them using the least resources and energy. You could say most pre-industrial buildings had some kind of "green building" traits, more or less. But what made us drift away from this fine tradition and end up on nature's opposite side? And what got us to realize again that the natural world is so crucial to us?

02-1 The Game Between Humans and Nature

Back in preindustrial times, humans had pretty limited power to reshape the environment. It's less that humans chose to roll with nature and more that they had no choice but to go along. Folks could only work at sunrise and rest at sunset, their production and daily life tied tight to nature's limits.

In 1765, the Brit James Watt improved the steam engine, kicking off the Industrial Revolution. A wave of tech breakthroughs followed, letting machines take over work that used to lean on human muscle, animal power, water, or wind. To people back then, the benefits of machines were obvious: Factories didn't need to hug rivers for water-power, electric lights and gear kept production lines humming 24/7, and man-made glass greenhouses kept plants growing through cold winters. During this stretch, human productivity shot way up, like we'd found a golden key to break free from nature's grip, ready to take on the wild and call the shots for our lives.

At the same time, tech progress pulled more folks off farms and into skilled trades, with people starting to pile into cities. Cities swelled fast, and urban buildings changed in materials and size: Elevators made tall buildings possible, while industrial materials like glass and steel swapped out stone and wood, making rooms bigger, windows wider, and construction quicker.

Artificial lighting, mechanical ventilation, and hot–cold AC were used more and more in buildings, letting their designs break free from local climate and natural resources, chasing a centralized, efficient, fully mechanized ideal instead. This shift also brought a whole new sense of what looks good.

But while folks were embracing industrialization and machine-made beauty, the Industrial Revolution's "side effects" started creeping in quietly. City population density kept climbing, and with it came a slow rot in the urban environment.

In Britain's Industrial Revolution, coal was the main juice powering those steam engines. Back then, folks didn't have a clue about protecting the environment—some even saw the thick smoke from burning coal, loaded with sulfur dioxide and toxic junk, as a steam engine's signature and a badge of the Industrial Revolution.

The French traveler Defoe once described Britain's iron-smelting hub Sheffield like this: "Sheffield's one of the dirtiest, smokiest cities I've ever seen. With small blacksmith shops missing tall chimneys, and the city full of hills, that smoke just spilled straight onto the streets. So people kept sucking dust into their lungs nonstop. Stay around the city long enough, and you're bound to breathe in coal smoke, letting it pile up in your lungs and destroying your health." In London, smoke mixed with fog, cooking up a thick yellow smog that hung over the city year after year. The novelist Dickens slammed it as "London's special flavor."

On top of that, river pollution worsened. In Britain's Industrial Revolution, the cotton textile trade was the first to go mechanical, pumping out massive amounts of pollution. Many textile mills dumped industrial wastewater straight into rivers, wreaking havoc on the environment and water. You might not guess it, but London's famous Thames River was once hit hard by industrial sewage. 1858 was the Thames's "Great Stink" year. That June, the river stank, and people in the Parliament building by the bank had to hang sheets soaked in disinfectants over the windows to block it out. In 1878, the *Princess Alice* pleasure boat sank on the Thames, killing 640 people—many didn't drown but died from swallowing the polluted water.

By 1950, Britain's urbanization hit 79%; other Western countries jumped on the high-urbanization train too. The fast urbanization march couldn't help but spawn problems like pollution, crowding, and city crime—folks back then called it "urban disease." In the 1950s, pollution problems appeared constantly in industrial hot spots. Some big ones, like London's "killer fog" and Japan's "Minamata disease" left many people sick or dead. Folks had to start rethinking the mess that comes from ignoring the environment.

London's Crystal Palace, the showcase hall for the first World Expo in 1851, was one of 19th-century Britain's architectural wonders and a major symbol of the Industrial Revolution era. It was the world's first big building made of metal and glass, put together onsite from prefab standard parts after being built offsite. The finished look was clear and simple, no fancy clutter, standing out sharp against the West's traditional building styles back then, grabbing tons of attention and praise right away.

In 1962, American marine biologist Rachel Carson's book *Silent Spring* came out, showing the public for the first time how pesticide overuse sets off a chain of environmental damage that ends up biting humans right back. From then on, pollution problems hit the public's radar for the first time, sparking wide attention and talk. Former U.S. vice president Albert Arnold Gore Jr. put it this way about *Silent Spring*: "Without this book, the environmental movement might've been held up a long time, or maybe not even started yet."

Sad to say, after the 1970s, environmental troubles didn't go away—they started getting bigger and trickier. Folks figured out that environmental issues weren't just one-off, local deals in rich countries anymore; they'd turned into global headaches like climate change and energy crises. Whether it's rich nations or developing ones, facing a global crisis like this, nobody can sit pretty on their own. People started looking at the

London street scenes under a blanket of pollution

Rachel Carson and *Silent Spring*

big picture, wrestling with the clash between population growth, economic push, and limited natural resources. In 1972, the United Nations Conference on the Human Environment kicked off in Stockholm, Sweden. It was the United Nation's first big meeting on environmental concerns, sparking the creation of the UN Environment Programme and setting the groundwork for global green teamwork down the road.

You Know What?

In 1987, the World Commission on Environment and Development dropped the report Our Common Future, rolling out the idea of "sustainable development": "Growth that meets today's needs without disrupting tomorrow's folks' ability to meet theirs." It tied together economic growth, social progress, and environmental care, breaking the hard link between human progress and running out of resources.

After the second UN Conference on Human Settlement in 1996, the ideas and rules of sustainable development were brought into architecture, weaving its environmental, economic, and social layers into building design and rating systems. Compared to the early "green building" idea, "sustainable architecture" zeros in on sustainability—it's not just about energy saving and cutting emissions but also how buildings touch the environment, economy, and society impacts, looking at buildings' long-term effects with a growth-minded eye.

Today, *green building* might be the term folks know best, and I will keep using it in this book to keep things easy to read. A building's sustainability across environmental, economic, and social layers is the big yardstick for whether it's a green win—and it'll come up again and again here.

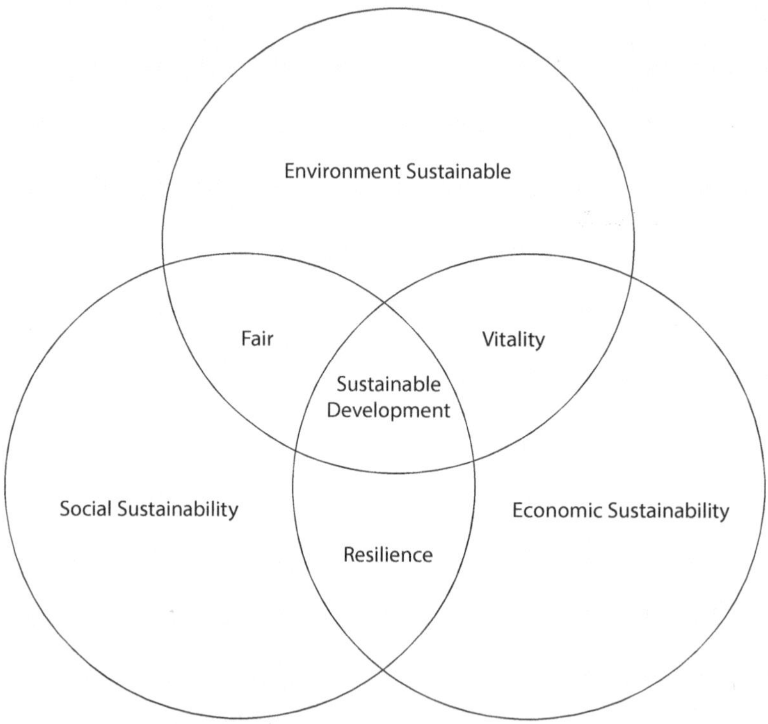

The Three Layers of Sustainable Development

02-2 Self-Sufficient Row Houses

Ever since environmental protection ideas started creeping into architecture, folks began taking a fresh look at how energy, resources, and human doings in building and using houses disrupt nature. How do you cut down on natural resources and fossil fuel use from a building angle, dialing back the harm to the environment? British architects Robert and Brenda Vale kicked off in the 1960s, tinkering and breaking new ground in green building, hands-down pioneers in the field.

In 1975, the Vales put out The Autonomous House: Design and Planning for Self-Sufficiency, laying out how to borrow local know-how, mix it with science, and build houses that run their own energy, treat the environment nice, stay easy to keep up, and carry a regional flavor. That book was translated into five languages, spreading far across the globe. It's seen as a cornerstone for modern green building.

The Vales drew inspiration from old-school building strategies, digging into designs that answer to local weather. This "passive design," which respects nature and uses resources smartly, is the one principle they stuck to and prized most. Next, let's take a peek at how architects handle the tie between buildings and the environment with real ingenuity.

You Know What?

Passive design's about tweaking a building itself to keep outside cold or heat comfy for folks inside, all while barely sipping energy. Put simply, it's using little or no man-made heating, cooling, lighting, or ventilation equipment, tapping sun, wind, and water smartly to give folks a cozy building space.

Baruch Givoni, an early passive design booster and co-founder of the Association of Passive Low Energy Architecture, summed it up like this: "Passive design covers **building design and material picks**, aiming to keep things comfy while cutting energy use." A building's design and material traits shape how much solar radiation it soaks up or lets through, plus indoor and surface temps, airflow, and vapor pressure. That way, they also shape how a building answers to the climate around it." So, passive design strategies swing in these directions:

- **Tweaking solar radiation:** soaking up or storing solar heat through glass or heat-holding material when it's cold; blocking it with shade when it's hot; keeping heat in or out with insulation.

- **Natural lighting:** letting in just the right sunlight for illumination through smart design.
- **Natural ventilation:** designing indoor–outdoor air pressure differences to get decent airflow and freshening.
- **Water vapor cooling:** cutting building heat in the hot times with water evaporating off the outside; tweaking indoor humidity with materials that soak up or let off moisture.

Broadly speaking, any approach that skips energy and resource use while making a comfy building space counts as passive design thinking. Can you think of any other passive design examples?

In a Nottinghamshire village called Hockerton, there's a row of houses that don't look like much at first glance. Built in 1998, this housing project sits next to a man-made lake, made up of five single-story units side by side, each with its own living space and outdoor garden. The project's owners wanted to show off and live out a building style and way of life that's light on the environment, self-sufficient, energy-smart, and full of everyday charm. Hockerton's houses might look plain and low-key, but they're anything but ordinary. It's the United Kingdom's first sustainable row house project and was listed by Guinness World Records as the world's most energy-saving home at the time. It was designed by none other than the green building trailblazers, the Vale couple.

The Vales put a big focus on how buildings answer to climate and the environment, plus making smart use of natural resources and energy. To hit that goal, they had to dig deep with early research and fieldwork, getting a solid grip on the local environment and weather.

The Nottingham area in Britain is pretty flat, with a mild climate. Summer highs hardly ever hit 30°C, and winter gets short spells of snow, but most of the time, temps stay above 0°C.

But winters here are often dreary and rainy, with not much sunlight. You can see the big challenge for architects was keeping the indoors warm and cozy through those gloomy winters.

So, the Vales rolled out these passive design strategies in the project:

- **Heavy build:** The project's main frame uses wood teamed up with concrete blocks. The north walls and roof were covered with earth, making a half-buried setup like a cave dwelling. Those thick walls and earth layers achieve the same cave-dwelling effect—warm in winter, cool in summer.

Hockerton Housing from above

- **Solar heating:** Big glass windows on the south side set up a suntrap room that soaks up solar heat easily. Plus, the house's angle and windows were figured out just right, letting the winter sun reach the deepest rooms while keeping it out in summer. Solar rays heat up the indoor concrete too, which acts like a heat bank, storing it up and letting it out after sundown to keep the place steady and comfy.
- **Insulation:** The house's floors, walls, and roof all received thick insulation materials, like wrapping the building in a hefty coat, cutting down outdoor heat sneaking in during summer and indoor heat slipping out in winter. The windows incorporate double- or triple-hollow glass, effectively maintaining thermal insulation as well.
- **User-generated heat:** In winter, in addition to the solar radiation absorbed by the glass sunroom, the operation of indoor appliances and human activity also produce considerable heat. All that warmth can pitch in as a heat source for indoor heating, bumping up the temperature.
- **Natural lighting:** The south-side sunroom doesn't just warm things up in winter—it brings in some great natural light too. Top skylights boost that natural lighting effect even more. Electric lights get used a whole lot less.

With this lineup of passive designs, the indoor temperature still hits around 17°C in the coldest winters, and the building's energy use drops to less than a third of what local regular buildings gulp down.

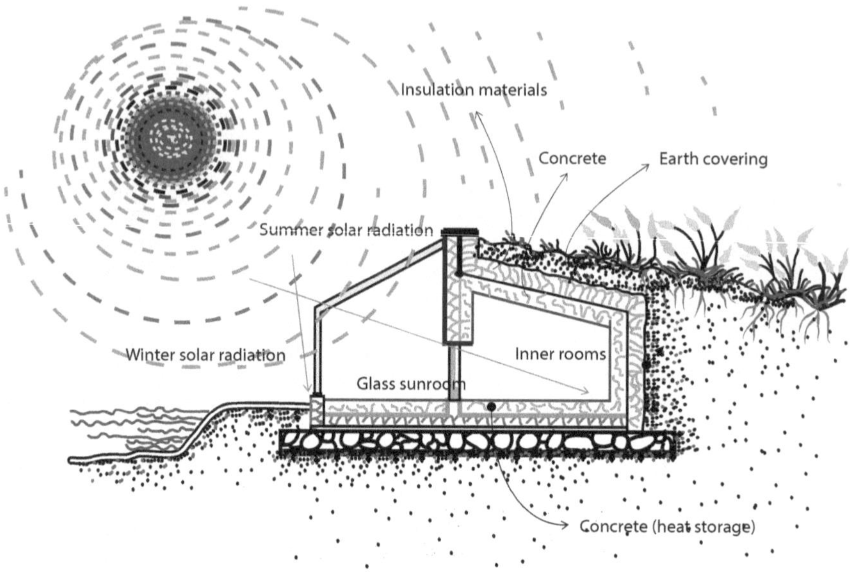

Hockerton Housing's passive design strategies (1)

Of course, passive design's knack for tweaking indoor conditions has its limits. For instance, regardless of how effective the natural lighting is, artificial lights must still be activated after sunset. Or, say, when daylight's too short or solar radiation's too weak, passive solar heating might not keep the indoors comfy enough, and you'll need some man-made heating equipment to pick up the slack.

But that doesn't mean passive design is useless. Passive design can cut down the time you need man-made gear to a minimum, slashing the energy use and emissions that come with it. That's where the saying "passive first, active fine-tuning" comes from. Here, passive means passive design, while active points to extra man-made systems. This saying sums up how they work together: Nail passive design first and then pick the right man-made equipment to back it up, letting both do their thing and play off each other, so you cut the building's harm to the environment way down while keeping folks comfy.

Hockerton Housing threw in mechanical ventilation and man-made heating systems as backups too. To power that equipment, they stuck solar photovoltaic panels on the roof and set up wind turbines around the site. So, energy-wise, this really is a "self-sufficient" building.

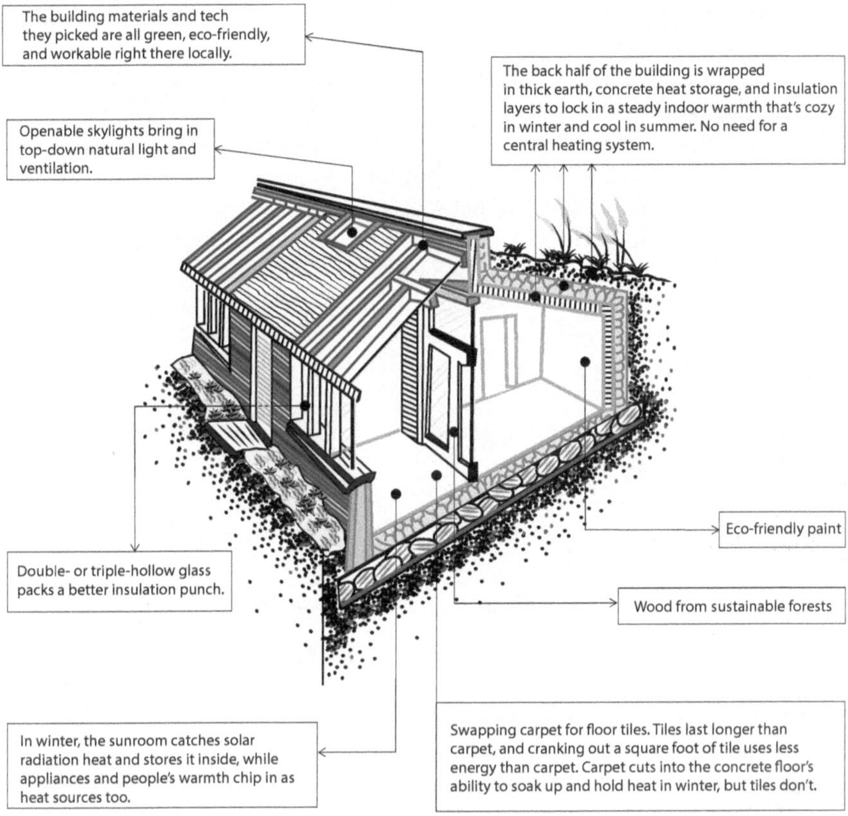

Hockerton Housing's passive design strategies (2)

You Know What?

Tuning solar radiation is a big deal for tweaking and steadying indoor temps. Next, I'll walk you through three passive design strategies tied to solar radiation heat:

1. **Thermal mass:** It uses materials with a big heat-holding capacity. When sunlight is strong during the day, thermal mass soaks up and stores the solar radiation heat. Come night, when the sun sets and temps fall, thermal mass lets out the heat it's been holding. This in-and-out rhythm slows down indoor heat loss, making the temperature cozier and steadier. Common thermal mass materials are dense and hefty, like stone, concrete, rammed earth, water, and the like.

2. **Greenhouse:** Also called a passive solar room, it taps the greenhouse effect glass creates. When the sun hits the glass, most short-wave radiation slips through to the indoors. But the long-wave radiation bounces off heated indoor floors, and air struggles to get back out through the glass. After a while, the temperature inside the glass room climbs higher and higher, turning into a "greenhouse."
3. **Thermal insulation:** Just like it sounds, its job is to block heat from moving through. Like a quilt, we wrap it around us in winter to stay warm, and in summer, popsicle vendors drape it over their goods to keep them cold. Insulation materials are usually light and full of holes, like foam plastic, lightweight porous boards, fibers, and materials like that. Hollow panels with air or inert gas layers count as insulation too.

These three strategies often get teamed up to keep indoor temps steady under all kinds of conditions.

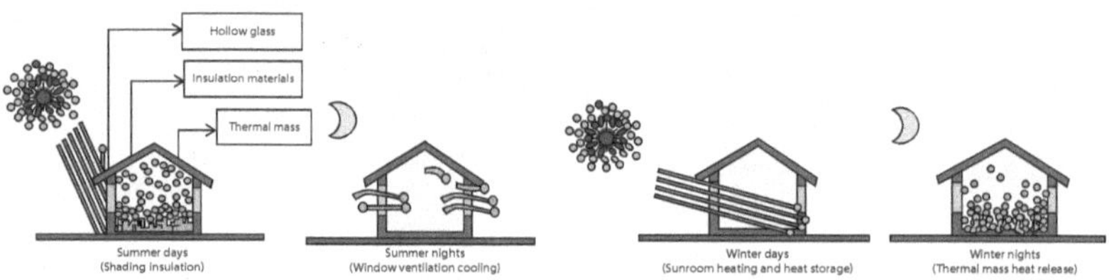

Hockerton Housing worked to be as self-sufficient as it could with water and food too. The project's drinking water comes from rainwater gathered off sloped roofs. After it is run through a filtration system, the water hits a high cleanliness standard. Non-drinking water mostly comes from rain collected off roads and fields, which is stored in tanks and then used after settling and a quick filter. For watering vegetables, which don't need top-notch water purity, they use raw natural water right from beside the house.

The sewage the building spits out gets handled right there too, meeting discharge standards. The key to that cleanup is the reed beds in the man-made lake. With smart figuring and design, sewage from all five houses flows into the lake, clearing out solid waste on one side while floating plants in the cleanup zone purify the water.

After about three months, the cleaned-up water hits a standard good enough to dump straight into rivers. Along the way, the man-made lake doesn't just clean the wastewater—it gives nearby plants and animals a place to live and the nutrients they need to grow.

At the same time, the Hockerton folks regularly scoop out the solid waste that settles from the sewage, turning it into compost for growing

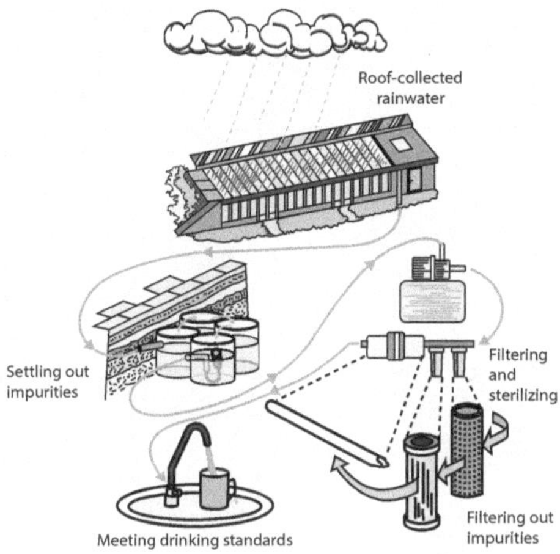

How drinking water is collected and treated

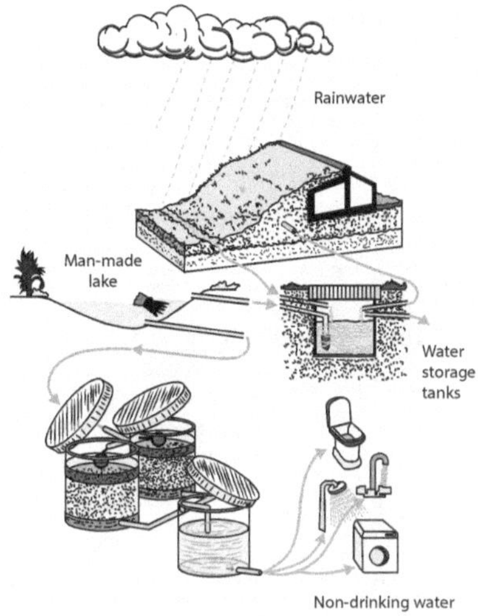

How non-drinking water is collected and treated

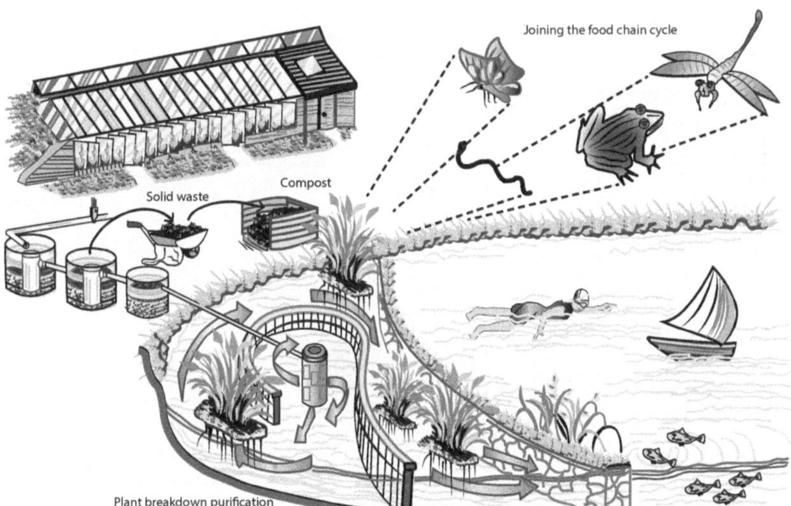

How non-drinking water is collected and treated

fruits and vegetables. They also raise chickens, sheep, and bees. About two-thirds of the vegetables, fruits, and eggs here come from their own hands. This nature-mimicking way of cycling resources cuts pollution from growing food and the energy and mess from hauling it long distances—it's not just green; it's also super energy-smart, saving residents over 60% on yearly building energy costs.

You Know What?

The environmental side of sustainable development is all about sorting out the tie between human growth and the natural world. To boost a building's environmental sustainability, you have to keep its harm to the environment as low as possible, maybe even figure out ways to do some good. Here's how you can tackle the design from a few angles:

- Buildings need to protect or boost things like the size of natural habitats, biodiversity, and productivity.
- Make sure that using natural resources over the building's whole life doesn't hurt the environment.
- Keep the trash and emissions from the building's full life from disrupting the natural world.
- Make sure water use and runoff over the building's life don't throw ecosystems off-balance.
- Keep the energy production and use over the building's life from knocking ecosystems out of whack.

Can you spot some sustainable design examples in the buildings around you? If not, what do you reckon could use some fixing?

The project's got a smart monitoring and evaluation setup too, pulling basic data to check the building's energy use, comfort, and concerns like that. Folks living there can dig deeper into how their daily habits tie to the building's energy and comfort, so they can pick smarter ways to use it better.

On the operations side, the project runs as a nonprofit co-op. Every homeowner, as a co-op member, has to pitch in 300 hours of volunteer work a year, handling concerns like the upkeep of the buildings, infrastructure, energy, water systems, and food growing. Plus, the co-op pays residents for another 300 hours a year to support gigs like eco-tourism, education events, and green building consulting. This community teamwork setup boosts everyone's know-how and skills for sustainable living, gets folks talking and working together, and lifts their sense of self and belonging, making them more eager to stick with this lifestyle and find joy and meaning in it.

You Know What?

Beyond sorting out its tie to the environment, a building's sustainability has two other big pieces:

1. Economic sustainability. It's about growing the economy without trashing the environment, hunting for a win–win way that boosts cash flow while keeping nature safe and sound. The main strategies include the following:

   - In building and upkeep, push for local jobs, solo start-ups, and small business growth.
   - Keep design and management as efficient as possible, holding high productivity to cut resource use, waste, and pollution.
   - Tap local know-how and skills big-time during construction and growth.
   - Track the building's impact on the environment, economy, and society with solid science during construction and growth, using that to steer things right.
   - Set up a fair, clear-cut economic policy and plan to give sustainable development steady juice and a safety net.
   - Back small-scale, local, diverse economies that fit the area's vibe.

2. Social sustainability. It's about making sure folks' physical and mental needs get met, lifting human growth so we can keep the sustainable development cycle spinning strong. The main strategies include the following:

   - The building setup needs to offer plenty of open public space, letting folks soak up and swap information as much as they can.
   - The building scenes must carve out lots of room and chances for education, including a focus on sustainable development learning.
   - The architectural setting should welcome all kinds of people while reflecting the spirit of the local region.
   - The building setup must keep people safe and help them stay healthy.
   - Building and running it needs plenty of input from regular people.

Hard to believe, right? These things that don't seem tied straight to nature are still quietly shaping how we see green building. It's only when folks realize green building won't bleed them dry but can pad their pockets and won't drag down their experience but make it cozier and richer that they'll take to this idea easily and keep pushing it forward.

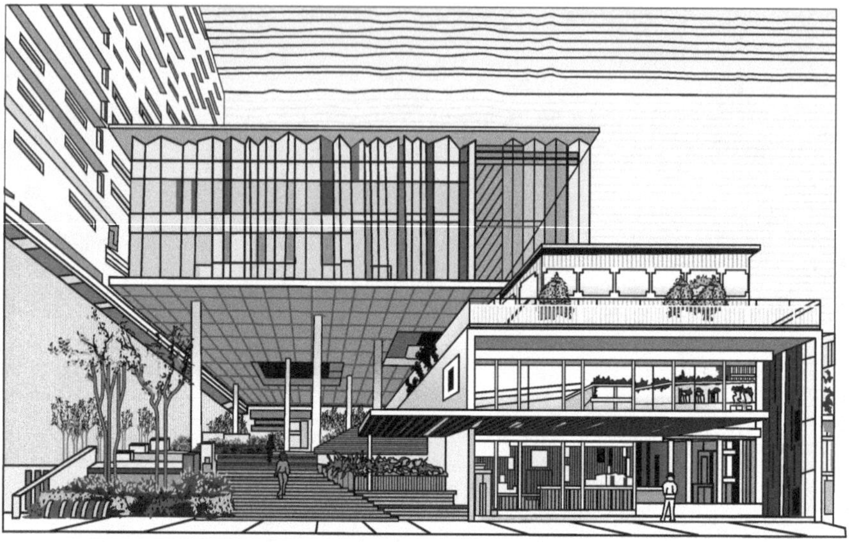

The Commons Commercial Community

02-3 Breezy Summer Commercial Community

After checking out a damp, cold British village in winter, next we're heading to a place with no winter all year—Bangkok, Thailand—to see a mall with no air conditioning. In different climates, architects stick to passive design that answers to local weather, but the nitty-gritty tricks change up a bit. Dancing with nature comes with all sorts of different steps.

Bangkok has a tropical monsoon climate, with average temps pretty much always above 25°C year-round, sometimes topping 35°C in summer. The air is humid, and sunlight is plentiful—it's like "summer all four seasons." In a climate like this, air conditioning feels like a must-have building gadget, and the energy it eats up and the electric bills it racks up every year aren't small. Going without AC all year in a climate like this—can it work? Sounds like one heck of a challenge!

Bangkok's The Commons is a commercial community, packing in a bunch of creative restaurants and shops. Cooking and eating in restaurants throw off heaps of heat, so going totally AC-free seems nuts, but The Commons pulls it off in public spaces outside the shops all year, crafting a warm, comfy, buzzing open spot for the city.

The Commons Commercial Community was built in 2016, a four-story commercial complex. It's not some high-end, big-budget "fancy"

Sunken courtyard

project—nowhere near a light rail stop or bustling business district, just sitting on a quiet community street in Bangkok. To keep costs down and hunt for a fresh take on community commercial complexes, architects from Thailand's Department of Architecture boldly pitched no AC in the mall's whole public area, as well as no escalators either.

To tackle that challenge, passive design is the go-to move you can't skip. In Bangkok's climate, the big questions are how to cut down excess solar heat sneaking in, how to boost airflow to chase off muggy warmth, and how to trim extra heat from daytime lights. So the best passive design strategies for the spot boil down to shading, natural ventilation, and natural lighting.

The architects kicked off by propping up the third and fourth floors with tall pillars, carving out a big shady patch on the ground level. In the scorching summer heat, folks passing by get drawn to this huge cool space, stepping in and feeling a quick blast of relief right away. Remember those cave dwellings—warm in winter, cool in summer? The Commons' sunken ground-floor courtyard pulls off something like that—air cools down there, with layered steps and chill spots dotted with plants pulling folks in to check it out.

Usually, commercial buildings have their outside faces plastered with flashy ads, and inside, light mostly comes from man-made bulbs. Sure, that pumps up the commercial buzz, but the building misses out on natural light and air, guzzling more energy for artificial lights and fans.

Heaps of mechanical equipment also throw off heat, needing even more juice to cool things down, kicking off a nasty loop. The Commons' design crew flipped the script, slapping big stretches of glass on the third and fourth floors' main faces to pull in maximum natural light and cut back on artificial bulbs. Outside that glass curtain wall, they added a faint metal mesh layer—keeps the building's face clean and sharp-looking, on one hand, while throwing some shade, on the other.

You Know What?

Beyond the thermal mass, sunrooms, and insulation we've covered, there's another big-deal strategy tied to solar heat—shading—which means using building parts or fittings to dial back solar radiation heat.

Common shading moves include stretching out roof overhangs, tossing up covered outdoor walkways, sticking shade pieces on windows, or leaning on nearby buildings or trees to block it out. The shape and size of shading bits hinge on the local climate, the building's angle, and the sun's height in the sky, among other factors.

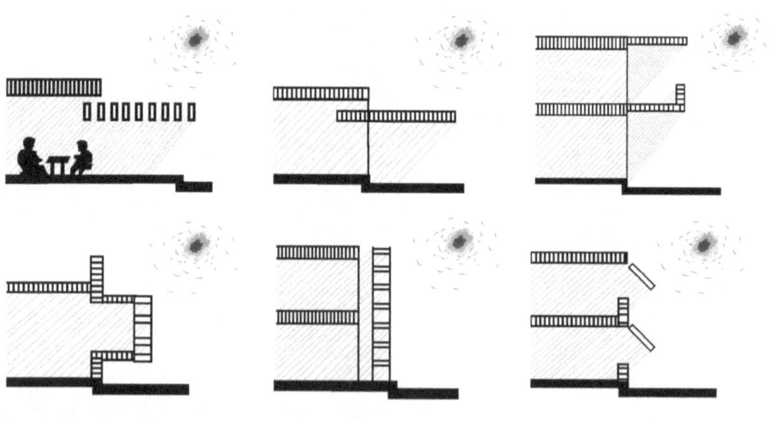

Some shading building part designs

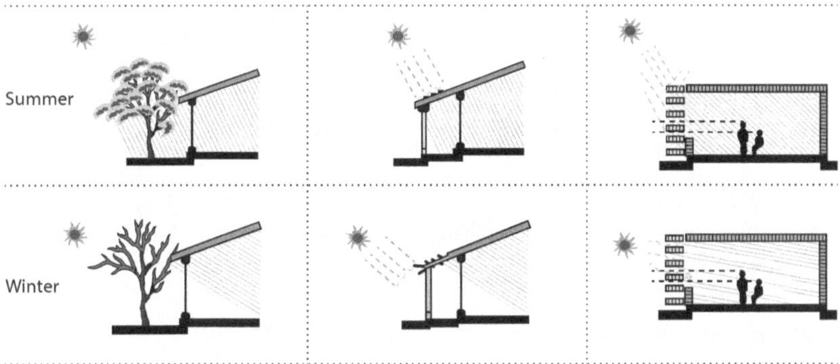

Shading features tie to seasons and solar angles: blocking summer sun, letting in winter sun

Courtyard space

The Commons' courtyard is a vertical, top-to-bottom space, wrapped with crisscrossing stairs and platforms. This setup pulls off a fun, flexible flow for moving up and down, guiding folks to different floors easily while also acting like a natural ventilation "chimney" that sucks air from the bottom up and out the top.

To crank up that chimney effect, architects stuck two big industrial exhaust fans at the courtyard's top. Outdoor hot air slips in through the shady ground level, cools off, and then flows into each floor via the

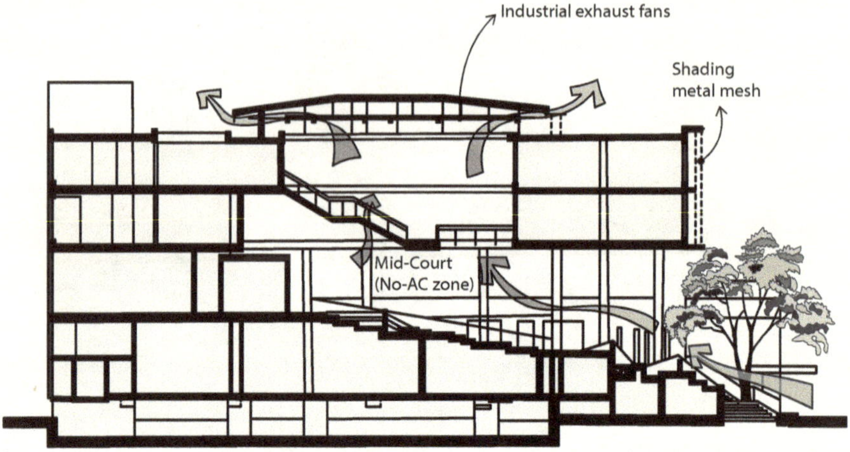

Ventilation cooling design strategies

You Know What?

Decent natural ventilation can speed up evaporation, whisk away extra heat, and keep indoor air fresh—it's a must-have piece of passive design in buildings.

Airflow's got a pattern—it always moves from high pressure or density spots to lower ones. So, the following are the two most common natural ventilation strategies:

Cross ventilation: In a windy spot, the windward and leeward sides of a building have different air pressures. Pop windows or holes on those different-pressure faces, and air flows from the high-pressure side to the low, kicking up cross-ventilation.

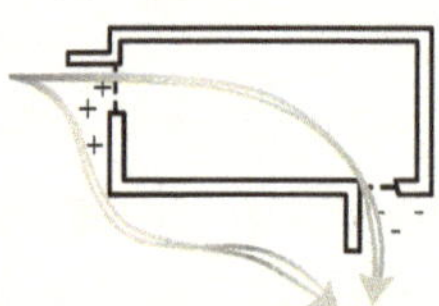

Cross-ventilation (floor plan)

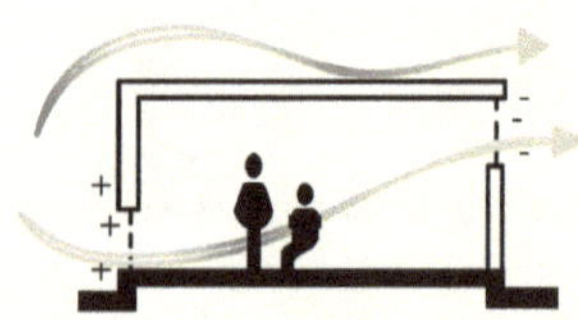

Cross-ventilation (section)

Chimney-effect ventilation: Hot air is lighter than cold air, so it rises. In a vertical space with windows or holes at the top and bottom, you get a chimney effect, pulling natural ventilation up from below. If we paint

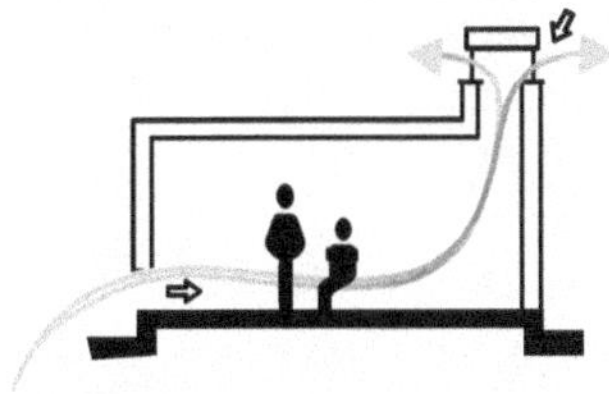

Chimney-effect ventilation

the "chimney" top dark, the air up there heats up more, improving the chimney effect.

Once you get how air moves, you can smartly tweak a building's angle and window or hole spots to whip up cross-ventilation. Like setting window faces at an angle to summer's main winds, popping multiple windows on different walls, opening holes at the bottom and top of walls, or tossing up wind towers to boost the chimney effect.

courtyard. Heat from building equipment and people warms the air, letting it rise slowly and get pushed out the top, pulling off great ventilation and cooling. Folks in the courtyard feel breezes wafting through, cool and comfy—no stuffy heat, no dry chill from the AC either; plus, it skips the hefty public space cooling bills.

See, The Commons is a mall, but it's also a community hangout—a spot to stroll, chill, listen to each other, and soak up natural light and breezes. It has loads of half-outdoor spots for activities—parents with kids, young folks meeting pals, or older people wanting quiet rest can all find their place here. Layered steps full of fun let folks wander through shops and spaces, climbing floors without even thinking about hunting for an elevator. Inside, folks can still peek out at the scenery, watch people coming and going, and feel the day's light shift—not stuck with the same old artificial lights and dazzling displays like in regular malls. Beyond daily shopping, there's a bunch of pop-up workshops, concerts, and little markets—cultural doings galore. It lives up to what the developers said: "Our intention is to build first a community, then a mall."

On the money side, The Commons nails low build costs and running fees. It's easy to see—skipping AC and elevators already saves a great deal on equipment costs and upkeep. The simple, bare-concrete look cuts out extra decorating costs too. It's not taken over by big malls, supermarkets, or chain stores—instead, it's packed with one-of-a-kind little eateries, craft shops, and kids' workshops. A mix of small businesses adds more fun and flavor to this commercial spot, even pulling in a steady stream of shoppers and sightseers. If you ever get to Bangkok, Thailand, wouldn't you want to swing by this community spot to feel it for yourself?

You can see green building doesn't just save energy and cut emissions—it's bigger than that, giving folks comfy, beautiful places to live, sparking teamwork, chats, and learning while bringing in cash and efficiency too. It's like a duet dance between humans and nature.

03 Green Building's "Positive Energy"

Green building, whether back in its early days or now, has one big job: "saving energy." At first, folks wanted to shrink building energy use as much as they could. But clever, challenge-hungry architects and engineers figured out they can't just drop a building's energy use to zero—they can make it churn out more energy than it eats up, flipping "negative energy" into "positive energy" to feed back into the community and nature. That way, a building stops being just a guzzling machine and turns into a gem that regenerates more energy and resources. How'd they pull that off?

03-1 A Building's "Lifetime"

If I told you a house could stick around forever once it's built, you'd probably call bunk on that. Obviously, buildings, like people, have a certain lifespan. From design to hauling and shaping materials to onsite building to running it to finally tearing it down—that's a building's "lifetime," what we call its "full life cycle."

When we want to size up if a building saves energy, cuts emissions, and runs cheap, we can't just eyeball how it works while it's up. Add up all the use, emissions, and payoffs over a building's full life cycle, and you might get results that catch you off-guard.

Remember that new rammed-earth farmhouse we talked about in rural Yunnan? Bet plenty of folks would say, "Building a brick-and-concrete house in the countryside's pretty cheap these days—cement, bricks, steel isn't pricey—so why bother with earth?" Well, let's peek at it through the full life-cycle lens and see which fits better in remote mountain areas: a new rammed-earth farmhouse or a brick-and-concrete one.

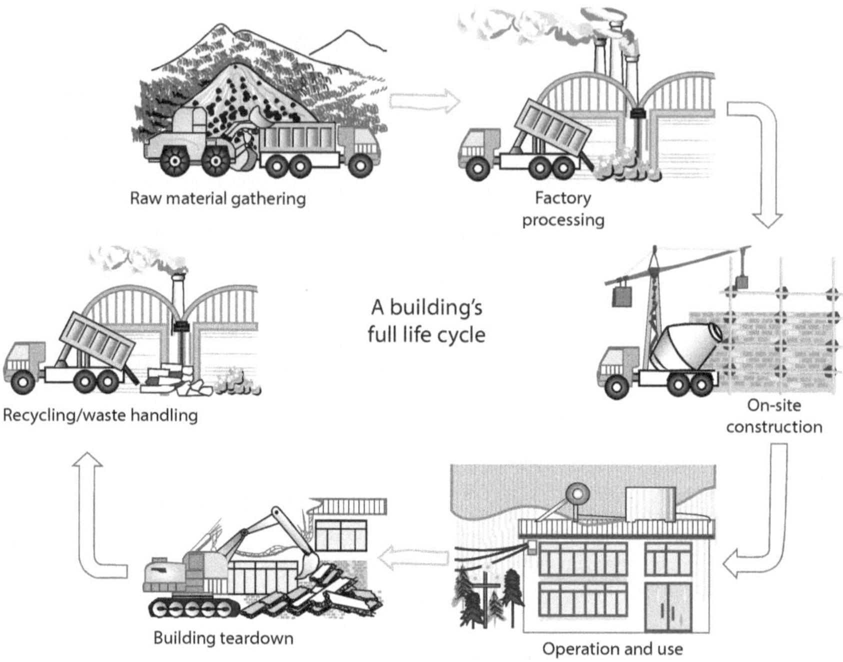

A building's full life cycle

The table later makes it clear: looking at the full life cycle, in rural spots with the right geography, climate, and a local rammed-earth building tradition, putting up a new rammed-earth farmhouse beats a regular brick-and-concrete one with less environmental drag, lower cash costs, and bigger community perks—meaning it's better for local sustainable growth.

So when we're talking about a building's energy use, first we've got to nail down which stage we're eyeing. If you want real sustainability, it's best to think about every chunk of that full life cycle, seeing them as one connected whole.

03-2 "Zero Energy" Housing Community

"Zero energy" housing—doesn't that sound a bit like a fairy tale at first? A big house holding all that living and working, guzzling energy every second—how could it possibly hit zero? Truth is, the British kicked off zero-energy building attempts back in the early 21st century. Built in

Comparing Guangming Village's New Rammed-Earth Farmhouse and Regular Brick-and-Concrete Farmhouse

	New Rammed-Earth Farmhouse	Regular Brick-and-Concrete Farmhouse
Material gathering	Main materials like raw earth, gravel, and sand can come straight from the spot. Environmental load: lower Economic cost: lower	Main materials like cement and steel need digging up from different places and hauling in. Environmental load: higher Economic cost: higher
Factory processing	Raw earth doesn't need firing—just simple physical work and it's ready. Environmental load: lower Economic cost: lower	Cement, bricks, and steel need high-heat firing and forging in factories. Environmental load: higher Economic cost: higher
On-site construction	Local villagers can handle the building after quick training. It can boost local villagers' cash, skills, confidence, and sense of belonging. Environmental load: lower Economic cost: lower Social benefits: higher	It takes outside know-how, hiring outside builders, and villagers footing the bill for most services. Environmental load: higher Economic cost: higher Social benefits: lower
Operation and use	1. Houses warm in winter and cool in summer can boost comfort, cutting down how long you need heating gear in winter and cooling equipment in summer. 2. Building gets done with local know-how, local materials, and local hands, and fixing it up or keeping it going can happen easily right there too. Environmental load: lower Economic cost: lower Social benefits: higher	1. Houses often lack decent insulation, so comfort is pretty low. To bump up comfort, you've got to crank up the heating in winter and cooling in summer for longer. 2. Fixing and upkeep might need materials, skills, or workers brought in from out of town. Environmental load: higher Economic cost: higher Social benefits: lower
Building teardown, recycling/waste handling	Raw earth materials are nontoxic and clean—break them down, and you can reuse them to build more earth houses or let them go back to nature. Environmental load: lower Economic cost: lower	Cement, bricks, steel, and materials like that take tricky steps to reuse, and tossing them as construction waste into landfills hits the environment some too. Environmental load: higher Economic cost: higher

You Know What?

Embodied energy—also called built-in or virtual energy—is all the juice it takes to dig up, process, make, store, and haul building materials.

Generally, materials needing a bunch of raw materials, tricky or energy-hungry processing—like high-heat firing, forging, or pressure—and made far from the building site carry higher embodied energy. On the flip side, materials found nearby with no intensive processing have lower embodied energy.

The University of Bath's Sustainable Energy Research Team in the United Kingdom rounded up a list of building materials' embodied energy (in MJ/kg)—it varies a bit by country and region, but the chart on the next page gives you a rough feel for how different materials stack up.

A 1996 UK study showed that embodied energy makes up about 20% of a building's total energy use over its full life cycle. The shorter a building's life, the bigger chunk embodied energy takes up. Cutting embodied energy is a job that needs all-around thinking. Sometimes, to make things last longer, we must use some higher embodied energy materials—then it's about weighing where, how much, and what they do; like the saying goes, "put good steel where it cuts."

Still, some buildings get torn down not because the materials wore out but because they don't fit the job anymore or just don't look sharp. So figuring out how to stretch a building's life with flexible insides or timeless design vibes is a design puzzle you can't brush off!

2002, the Beddington Zero Energy Development—BedZED—was one project that grabbed a ton of attention back then.

But I've got to level with you—early on in zero-energy building, most so-called zero energy meant zero-energy use while running, not over the whole life cycle, and BedZED is no exception. Don't feel letdown yet—even now, not just back then, hitting zero-energy use while a building is up and running is a pretty big deal. So how'd this project pull it off?

BedZED sits in London's suburbs, with 82 housing units and 2,500 square meters of office, studio, shop, and community service space. Designed by British architect Bill Dunster, who's all about green building, the project aimed to slash energy use and environmental harm to the bone, showing off an eco-sustainable zero-energy community.

To hit that goal, the project crew got to work right from picking the spot. BedZED's spot didn't gobble up fresh farmland or wild woods—instead, they picked a patch already built on but sitting idle, an old,

Embodied energy of different materials

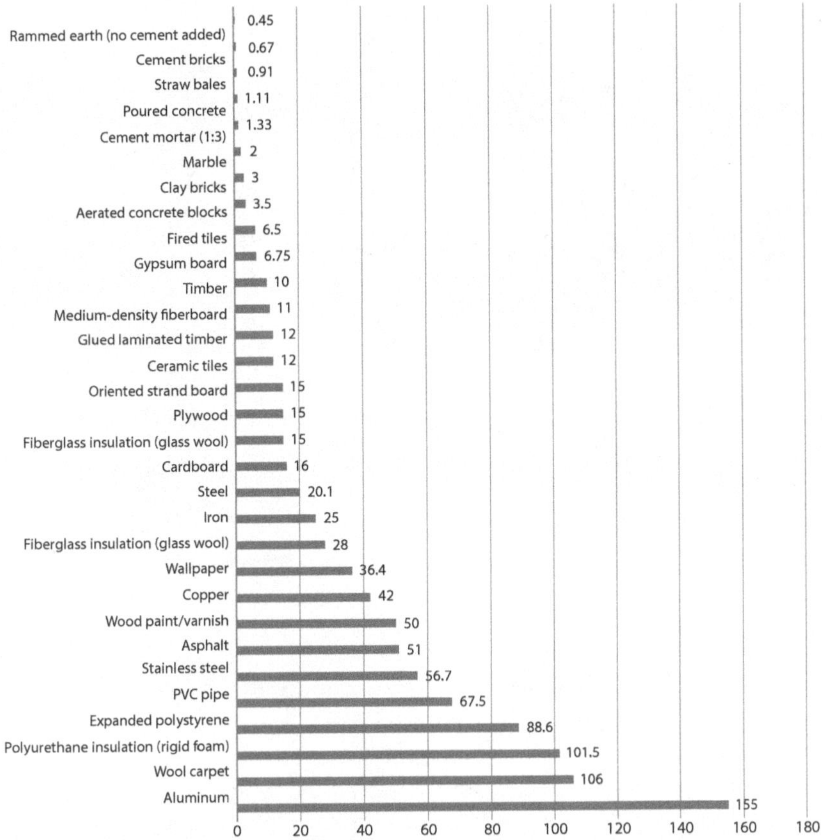

abandoned factory site. That kind of pick kept the natural ecosystems around the building from taking a big hit. Number crunchers back then figured that if you turned all the potential environmental good from this low-impact, high-quality build into cash, it'd be worth about £200,000. Laying it out in numbers like that showed off the project's strengths and won over the local community's backing.

To trim embodied energy consumption, BedZED leaned hard on recycled steel bits and softwood pegs from local torn-down buildings, plus a big haul of FSC[1]-certified timber. Most of the building materials

Beddington Zero Energy Development (BedZED)

and workers came from within 50 miles—about 80.5 kilometers—of the site. That didn't just slash the embodied energy of the materials—it also boosted local jobs and fattened workers' wallets too.

For how it works, BedZED isn't just a housing spot—it mixes living and working spaces in a "work–life combo" setup, nudging folks to work near home, cutting long commutes and the energy and emissions they pile up. On top of that, BedZED trimmed gas-car parking spots; added bike parking, free electric car chargers, and a car-sharing club; and tied in smoothly with all sorts of public transit around the community. Those moves seriously shrank how many miles folks drive.

[1] FSC, or forest certification, also called timber certification, is a market-driven tool to push sustainable forest management, hitting ecological, social, and economic goals. It works by independently checking forest operations, proving to buyers the wood is from sustainably run forests, and linking "green shoppers" with producers aiming to up their game and grab bigger market shares for better profits.

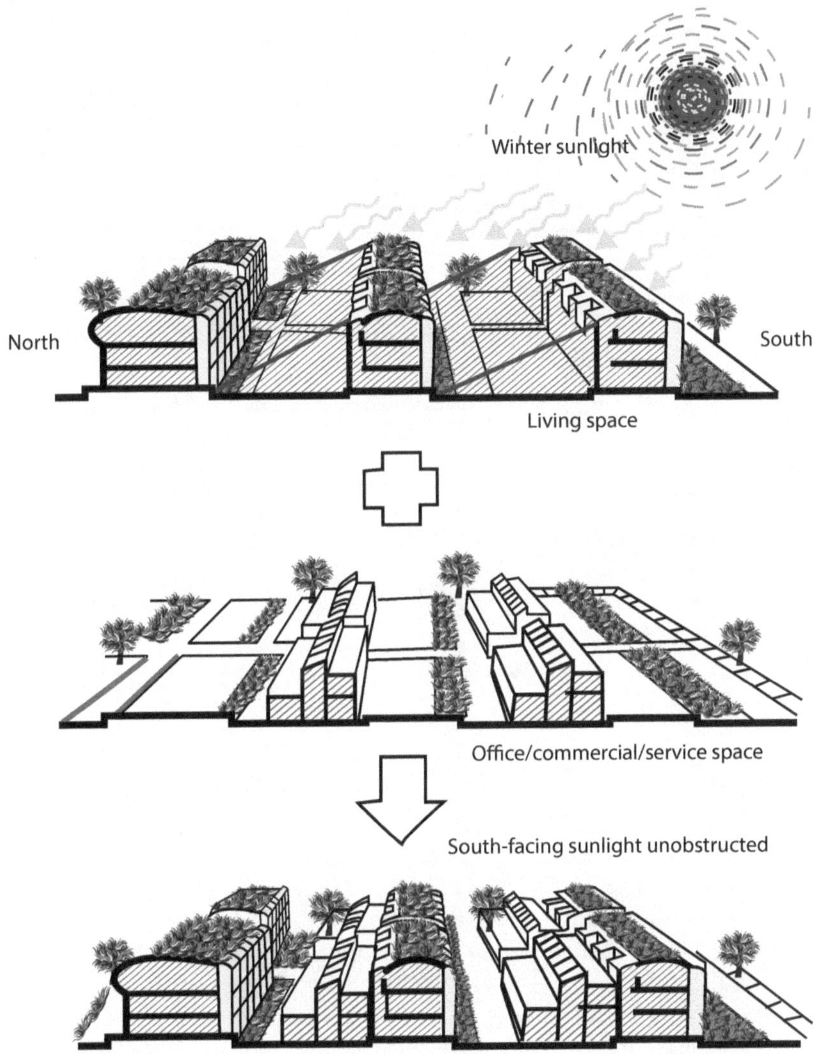

Building shape designed to match sunlight angles

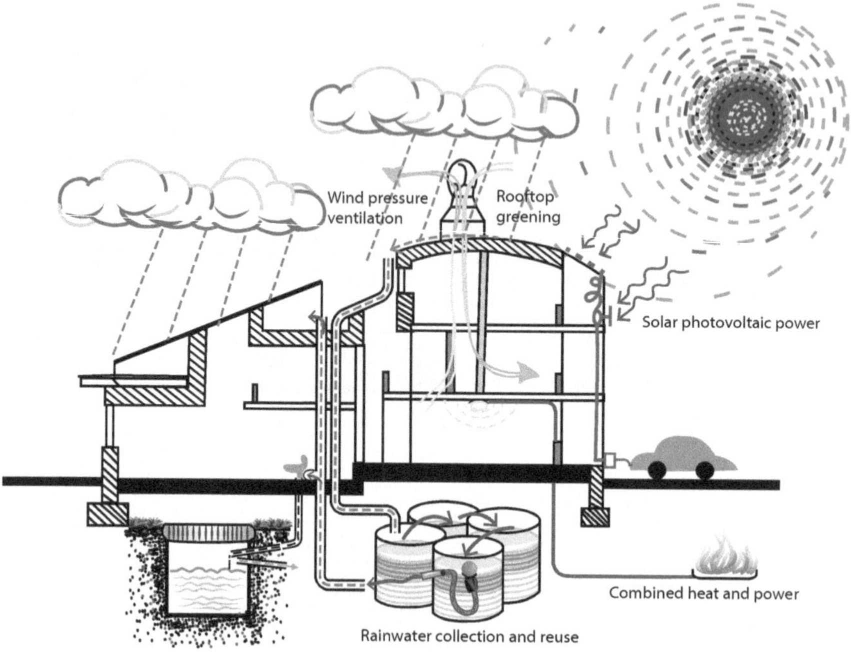

Blending passive design, renewable energy, and water reuse

Design-wise, BedZED, like other winning green buildings, answers local weather full-on with smart, effective passive design strategies. London's summers are short and mild, winters long, damp, and chilly. For that climate, the architect crew rolled out two big plays: passive solar heating and natural ventilation.

In winter, big south-facing glass windows can soak up slanting sunlight to heat the indoors, storing warmth in floor slabs and other thermal mass, letting it out slowly after sundown to keep the place cozy. Add in the heat from people moving around, appliances humming, and cooking, and it's enough to warm the house to a comfy human temp, cutting way down on the need for extra heating.

In summer, as the sun's angle shifts, sunlight gets blocked by outer walkways, not beaming straight inside—some light bounces through skylights a few times to brighten the place, hitting lighting needs without letting in too much heat. The roof has a 300-millimeter-thick rock wool insulation layer, keeping summer heat out too.

Plus, BedZED's building roofs, balconies, and courtyards have plenty of green space. Damp soil can hold heat, while plants sweat it out through transpiration, soaking up warmth. That helps lock in steady, comfy indoor temps even more. Those colorful standout pieces on the building roofs are passive wind pressure ventilation systems, improving the natural airflow inside.

Doesn't this focus on design feel familiar? Much like Hockerton Housing from Chapter 2, right? Both projects—one in Nottinghamshire, one in London's suburbs—aren't far apart, with close geography and climate, so it's no surprise their passive design ideas line up.

Solid passive design means BedZED's houses don't need extra heating gear—just power and hot water do the trick for daily life. So BedZED's main juice comes from combined heat and power—burning factory sawdust scraps to crank out electricity and hot water for the community. Sawdust is biomass energy, a renewable source, and reusing factory waste like this pulls off a bunch of wins at once.

Truth be told, BedZED's zero energy isn't really zero—people's daily living always burns some energy. Its zero energy means that no nonrenewable energy gets used in the community—all the juice comes from renewable materials made right there. That's hitting a kind of "self-sufficiency" in its own way.

You Know What?

- **Nonrenewable energy**—also called non-regenerating energy—is something that took billions of years to form in nature, can't bounce back quickly, and becomes scarcer with big use until it runs dry someday. Common everyday materials like oil, coal, and natural gas—those fossil fuels—are nonrenewable energy. Same deal—nuclear fuel, minerals, and such are nonrenewable resources.
- **Renewable energy**'s the stuff nature never runs out of—or that cycles back quickly in the short term. Solar, wind, tidal, and biomass energy—such as firewood or biogas—all count as renewable energy. Same goes for materials like timber and bamboo—grown big, cut down, and regrowing—those are renewable resources.

Monitoring data from 2003 show BedZED slashed 88% of the building's heating energy, 57% of hot water energy, and 25% of electricity use compared to the UK average back then—and all that power came from solar and combined heat systems. Plus, BedZED was the world's first "zero-carbon community." (What's zero carbon? I'll tease that here and spill the details in Chapter 4.) All in all, BedZED, as the United Kingdom's first big, dense zero-energy community, set an exceptional example, giving folks heaps of confidence to spread zero-energy building-wide. It even sparked a bunch of follow-up projects, like the Zero Carbon Pavilion at the 2010 Shanghai Expo.

Still, no green building nails perfection. Even the slickest designs hit plenty of unknowns when they're up and running. A 2010 survey showed that while BedZED folks and nearby neighbors were pretty happy with the community, the project's hiccups stood out too.

Back then, BedZED's combined heat and power system was pretty tricky tech-wise and cost a lot to keep up. With ongoing tech glitches, the combined heat and power setup became less and less cost-effective. After a bunch of twists and turns, post-2017, BedZED's combined heat and power could only churn out hot water for the community, no more electricity. And the biomass pellets it ran on stopped coming from local wood waste—it had switched to buying the pellets from Spain instead.

Looking back at this project, folks realized rigging up a small setup with everything from energy to water treatment to be fully self-contained isn't always the slickest way to go. Sustainable development needs bigger-picture planning and more teamwork tie-ins. A "go-it-alone," "utopian" green building as a demo is a solid start and try, but folks need to learn and rethink, pushing sustainable ideas to more levels and wider scopes, not just stopping here.

Every building project, over its decades or even centuries of life, can't dodge all sorts of changes and hurdles. We ought to look at it with eyes open to change and growth too. Taking change in stride and gearing up for challenges is a big skill that architects and building runners need to have. Like architect Bill Dunster said, "We never claimed this was a perfect project—no denying, there's room to step it up."

Nevertheless, smart material picks and passive design still shaved off hefty chunks of embodied and running energy for this project. Ongoing updates and tweaks kept the building ticking along smoothly too. Through the architects' design, the building still delivered a sweet, comfy indoor space on such low energy, winning over folks who use it and sparking more green builds—maybe that's BedZED's biggest win.

03-3 "Positive Energy" Green Factory

Is "zero energy" the ultimate dream for green building? Nope, not quite. Green building went from low energy to zero—what's next? Taiwan's O'right Green Factory not only sips tiny bits of energy but makes more than it uses, selling the extra to the local grid for others to tap. That's the "positive energy" building I want to show you.

O'right Green Factory sits in Longtan District, Taoyuan City, Taiwan—a plant pumping out organic daily goods. In September 2006, O'right factory bosses sat in an old meeting room, hashing out how to keep the plant and brand running sustainably. They wanted not just their products to be natural and green but the factory making them to live in maximum harmony with nature too. So they brought in architect Hung Ying-chin and his research crew, aiming to build a "green cosmetics factory" to flip people's wrong ideas about resources and show off a push for green, sustainable living.

After a year and a half of scouting and surveys, the team settled on a half-hillside spot in Longtan District, Taoyuan City, for the factory site. This spot has gorgeous surroundings, is full of life, and sits higher up, about 2°C cooler than city areas. It's not just loaded with sunlight—the terrain also makes northeast monsoons swirl back here, perfect for solar and wind power.

The factory's main frame uses bare concrete. The design crew beefed up the protective layer around the rebar in the concrete, stretching the building's life to the max and making that concrete use more "worth it." Plus, bare concrete buildings don't need outer wall dressing, skipping the energy from materials, labor, and upkeep that'd come with it.

To nail down the best passive design strategies for local weather, the design team spent a full two years gathering sunlight, wind, and climate data from the area. The big local climate hurdle is a hot, drawn-out summer—most regular buildings burn their running energy on cooling AC systems. So the building design has to dodge too much solar heat indoors while pulling in maximum natural light to cut down on artificial bulbs. The design crew stuck the main office spaces facing north, using big north-facing floor-to-ceiling glass as the prime light catcher. The south face has a more closed-off solid concrete wall, with just small, tucked-away ventilation windows. That angle also lines up with the local main winds, helping the building's cross ventilation.

Doesn't this south-facing-north setup flip what you thought about building angles? Truth is, the south-window habit isn't a one-size-fits-all fix—it's got to fit the local scene. More than blindly slapping on a formula, I'm more into this "here-and-now building way" that respects the site's reality, picking the best strategies for each spot's unique conditions and story.

Plus, the design crew leaned hard on natural ventilation and the simplest cooling equipment—ceiling fans—to chill the building inside. To keep indoor air moving freely, they used clear glass

O'right Green Factory

The building's south face

partitions between offices, leaving gaps at the top and bottom. Inside, they ditched all the usual office cubicle screens. The company boss says: "Everything's open and clear here, but the downside is it's noisy—folks talk loud!" Meanwhile, workers figure that without screens, it's easier to bond with colleagues and chat anytime, and that's a handy perk for work.

You Know What?

In the warm, wet tropical and subtropical monsoon climate, sometimes the temperature isn't super high, but the humidity's thick, making the weather feel muggy and uneasy. If you flip on the AC then, you've got to crank it lower to feel cool and dry, but that gets too chilly and burns more energy. Truth is, there's a simpler, slicker way—just use a fan.

These days in home decorating, slapping up ceiling or wall fans isn't so common anymore. But back when AC wasn't everywhere, fans were a must-have gadget in every home. When it's a bit stuffy, fans kick up airflow, evaporating sweat off your skin, which cools you down and makes you feel comfy. Compared to AC, fans use less energy, and setting them up or keeping them going is a breeze.

Sure, when it's blazing hot, fans can't fully swap out AC, but in that shift from mild to scorching, or at night when temps dip and folks are lying quietly, a fan's a solid pick. With green awareness picking up today, fans are sneaking back into folks' sights. There are ceiling fans mixed with lights, plus rechargeable cordless standing or desk fans that you can move around easily. Would you give a fan first dibs when it's just a tad muggy?

To pull off a building's "positive energy," equipment that churns out renewable energy is a must. Beyond tossing up wind turbines, O'right Green Factory's roof has 11 round solar panels of different sizes, tapping the local hefty sunlight to pump out nonstop power. Those solar panels tilt a bit south to catch the best sunlight. They're like 11 leaves basking in the sun, doing their "photosynthesis."

With this "all-hands-on-deck" design, O'right Green Factory only needs AC for less than two months a year, and just 6 hours a day at that.

The solar power system and wind turbines don't just cover all the plant's electricity—they've got extra juice to sell back to the power

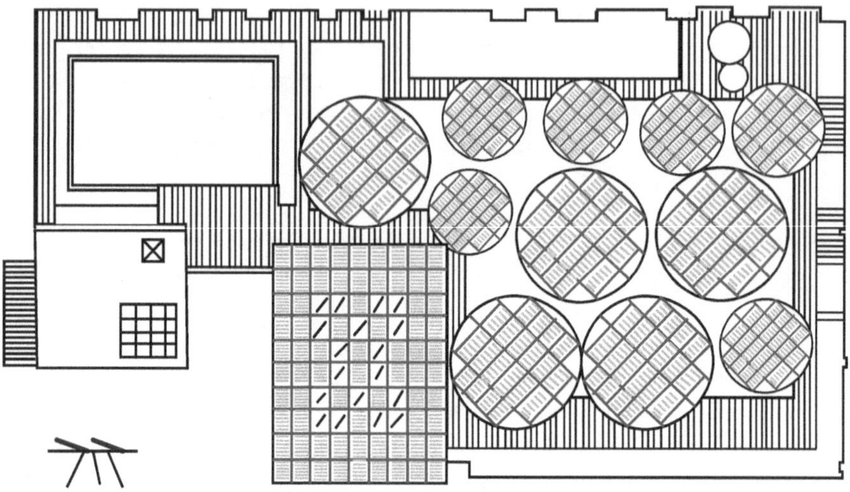

Photovoltaic panels raised off the bottom, tilted slightly for top power efficiency

Photovoltaic panels on the building roof

Photovoltaic power system

company as well, letting more folks tap renewable energy. In four years, O'right raked back 2.1 million New Taiwan dollars—about 470,000 RMB—in electric bills.

Of course, building a green factory like this didn't come cheap up front. Records show O'right Green Factory's floor space is 1,684.35 square meters, with a total build cost of 90 million New Taiwan dollars—around 20.691 million RMB. That breaks down to a per-square-meter cost of 12,284.26 RMB. But if the building's running-stage electric bills are all self-covered, no cash out of pocket, and you can sell extra power, then over the long haul, that big up-front cost evens out bit by bit.

See, green building isn't just high-tech and wild ideas—it's a ledger too. For owners or investors, it must be cheap and sharp—no one wants a money-losing gig. For the bigger environment—even the planet—you've got to tally up the energy and resources used, the pollution and greenhouse gases pumped out, and see if the good it does can balance it out slowly to hit "breakeven." We'll dig into this number-crunching talk in Chapter 7.

Worth noting, O'right Green Factory's site is out in the suburbs, already lush with plants and nature—a hangout for frogs, birds, and

You Know What?

Solar power systems can turn sunlight into electricity through semiconductor chips. Compared to other energy-making ways, it's got no moving parts, so no noise; no burning, so no fuel; and sunlight's pretty much endless and free to grab, making solar a green energy folks are liking more and more.

Solar power systems are split into two big types: standalone and grid-tied.

- **Standalone power**, like it says, is a solar system running solo for one building or a cluster of buildings. It uses batteries to stash power, banking extra juice from the solar system when sunlight is strong during the day and then tapping that stored power at night or when the weather dims the sun to keep the building humming.

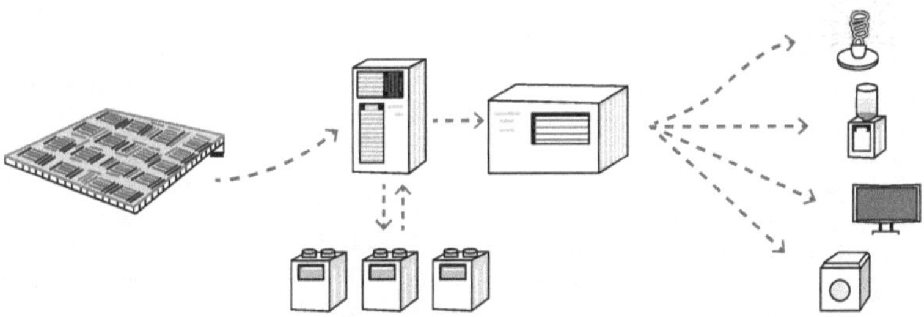

Standalone power

- **Grid-tied power** hooks the solar system into the city's power web, using the grid like a battery. The solar system feeds its own building first, and when it makes more than it uses, that extra power flows into the city grid. That way, it pulls off a "positive energy" move, selling clean power.

Of course, no system's picture-perfect—solar power setups have their limits too.

For one, sunlight is at the mercy of geography, climate, seasons, and weather. So you need energy storage systems to keep the power steady. That bumps up the tech challenge and the price tag too. On the flip side, while the total solar energy hitting Earth is huge, its density is pretty

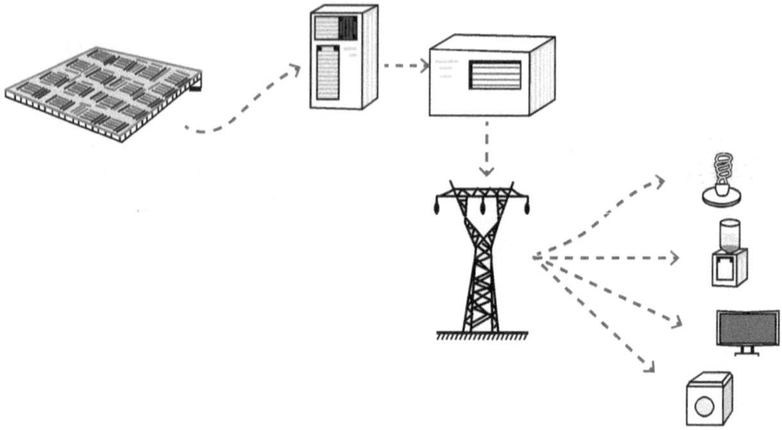

Grid-tied power

thin. So the gear turning sunlight into power needs a big enough spread to grab enough juice. That means building a solar system's cost climbs up pretty high too. Plus, making solar panels, batteries, and parts eats energy and resources—and kicks out some emissions along the way. You must manage emissions over the whole life cycle to really pull off eco-sustainability.

Still, tech's changing fast these days—by the time you read this, new strategies might have rolled out a few times, sorting those issues one by one. Like we keep saying, we have to look at things with eyes on the future. As tech marches on, folks will surely dig up more efficient, sustainable ways to power up.

lots of little critters. The architects put real thought into the landscaping too, keeping those animals' homes as undisturbed as they could.

Around the building, thick greenery—over 60 big trees and more than 10,000 plants—whips up a lively landscape, with heaps of native broadleaf trees kept in fine shape. The building's roof has green roofing and sky gardens, planted with local materials that do not need much human fussing. That greenery doesn't just cool the building and cut AC use—it gives birds and bugs a place to crash too. Rain hitting the roof first washes dust off the solar panels and then waters the roof plants, trickling through soil and pebble layers below to filter into a ground-level eco-pond. Water-cycling eco-ponds around the building stash

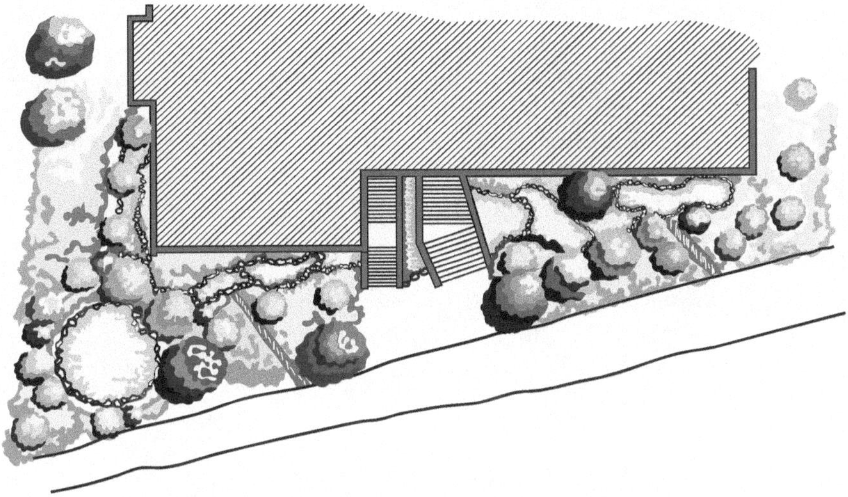

Eco-landscape design with ponds and plants around the building

recycled rain, clean it up, and offer a hangout for local bugs, small fish, frogs, birds—little critters—building a wild, fun ecosystem.

That way, the building doesn't just give humans a comfy work spot—it also looks out for plants and animals thriving too. And giving plants and animals a home flips back to save the building's water and energy use. Any way you slice it, that's a win–win deal pulling off a bunch at once!

You Know What?

Sure, humans call themselves "masters of all things," but our survival and growth can't break free from the Earth's ecosystems. On this blue planet, beyond humans, there are 8 million kinds of plants and animals. We lean on each other, weaving a giant web of life where every piece plays a big part.

"Biodiversity" means the variety of life on Earth—covering all species, the ecosystems they live in, and the genetic mix within and between them.

Biodiversity is the backbone holding up all life on land and underwater, and it touches every corner of human health too. It keeps the climate steady, guards water and soil, and holds up normal ecological cycles—doing huge favors for all of us humans. Sad to say, human doings have already shifted land and sea environments in ways we can't dodge. City sprawl and our grab for nature's goods keep shrinking plant and animal homes, while the carbon dioxide we pump out doesn't just mess with the climate—it's also souring the oceans, putting sea life on the ropes. Losing biodiversity is going to hit humans hard—think food and health systems falling apart.

So figuring out how to build human digs while keeping biodiversity harm as low as possible—or even boosting it in a spot—is a must-think-about issue for any architect worth their salt. We're not just crafting human living spots—we're also sharing this one-and-only Earth with every plant and critter out there.

04 The Tussle with "Carbon Emissions"

Beyond "green building," you've probably spotted "low-carbon building" or "zero-carbon building" popping up in the media plenty. So what's this "carbon" really about? How do buildings and carbon even get tangled up? This chapter is here to walk you through how architects wrestle with carbon emissions.

04-1 The "Culprit" of Climate Change

Remember how we talked about where green building ideas came from? Back in the mid- to late 20th century, industrial pollution woke folks up to how big a deal environmental protection is. So does fixing industrial pollution wrap up all the environmental woes? Nah, it's way trickier than we'd guess. As time rolled on, folks saw that just tackling pollution didn't cut it—a fresh wave of environmental messes hit wider and got knottier to figure out. Global climate change is one of those issues that hits wide and cuts deep.

Science has figured out that the concentration of greenhouse gases in Earth's atmosphere straight up affects global average temps. Since the Industrial Revolution, human activities have jacked up greenhouse gas levels quickly and steadily, pushing global average temps up too, at a pace way beyond Earth's natural climate shifts and life's evolution speed. As research keeps rolling, mainstream scientists reckon more and more evidence shows this "unnatural" climate change is real and packs a massive punch to ecosystems and human lives. The heaps of greenhouse gas emissions from humans'hustle are the "culprit" behind climate change.

You Know What?

When solar radiation hits Earth, some energy bounces back to space, while another chunk—mostly long-wave radiation—gets soaked up by the atmosphere and ping-pongs around inside it. The greenhouse effect is when a planet's atmosphere grabs solar radiation energy, bounces it around inside, and heats up the surface.

Gases in the atmosphere that can soak up long-wave radiation and "store" solar heat are called greenhouse gases—or GHGs. Water vapor, carbon dioxide, methane, nitrous oxide, and fluorinated gases are the big players among greenhouse gases.

Out of all greenhouse gases, carbon dioxide pitches in the most to the greenhouse effect, so it's the top dog to watch for tackling climate change. Folks use the weight of carbon dioxide emissions as the yardstick for greenhouse gas output—shortened to carbon emissions.

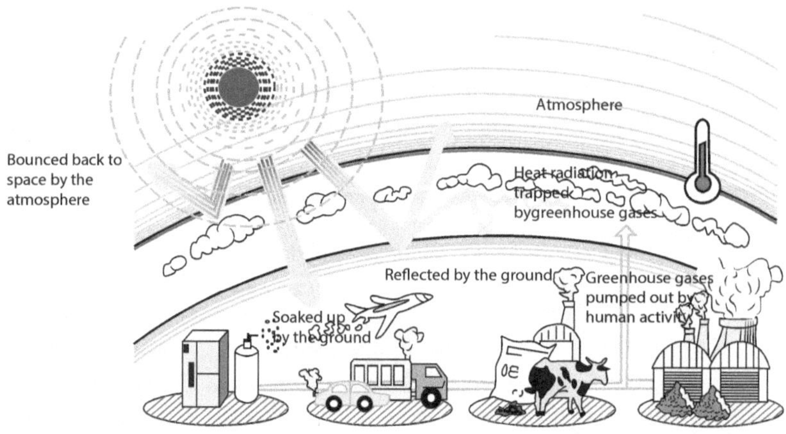

Greenhouse effect

Earth's greenhouse effect is a bit like the greenhouses we build for veggies or tropical plants—that's where the name comes from. And that passive design strategy we talked about earlier for green buildings—the sunroom—leans on the same idea. See, the same effect playing out in different spots can lead to totally opposite outcomes. Greenhouses are great for hothouse veggies, but Earth's greenhouse effect isn't so hot for us humans. A speeding climb in global average temps can warm up seawater, shrink ice and snow cover, lift sea levels, crank up wild weather—a whole string of global headaches, even disasters. Ecosystems and Earth's climate setup will tip out of whack too, seriously threatening plants, animals, and human survival.

In human doings, burning fossil fuels, farming, and industrial churning all pump out greenhouse gases. Since the Industrial Revolution kicked off, humans have bumped up atmospheric carbon dioxide levels by 48%. Cutting carbon emissions is the top job to rein in the greenhouse effect and slow global climate change.

The Intergovernmental Panel on Climate Change, set up by the World Meteorological Organization and the UN Environment Programme, pumps out regular reports to dish up solid science on climate change. Scientists figure if the climate warms 1.5°C, the globe's in for a bunch of nasty hits. Folks by the sea might lose the land they live off, more wild weather and disasters could slash food crops, shrink fish stocks, throw ecosystems off, and that chain reaction could sink more people into poverty or even put their lives on the line. If the climate warms 2°C, the bad effects could double, maybe even pile on more. For our survival and our kids' and grandkids' future, humans need to pull out all the stops to keep this from happening.

To cap global warming at 1.5°C, every country and industry has got to team up and slash carbon emissions across the board. By 2030, global human-caused net carbon dioxide emissions have to drop about 45% from 2010 levels, hitting "net zero" around 2050. This massive gig isn't something one country or region can tackle on its own—it takes the whole world pitching in. Sure, humans' hustle has roughed up the environment plenty, but the ones who can turn the tide and fix this mess might just be us humans too. It seems like humans have never quite grasped as deeply as today that Earth and its atmosphere are precious treasures we all share. Like it or not, our fates are tied up tight together now. That's why we've got to build up a lot of confidence and trust in each other; believe that humans, as a whole, can rethink and tweak ourselves; and believe that while you're busting your hump for a shared goal, others are too.

So how do we cut carbon emissions? First off, obviously, dial back or swap out human doings that spit out carbon dioxide, pushing for smarter, more responsible consuming and living ways. Next, because most energy we use now—coal, oil, natural gas, those fossil fuels—pumps out heaps of carbon dioxide when burned, swapping some of that for solar, wind, water, and tidal energy that doesn't cough up carbon is a key move to trim emissions.

Global carbon dioxide emissions by sector

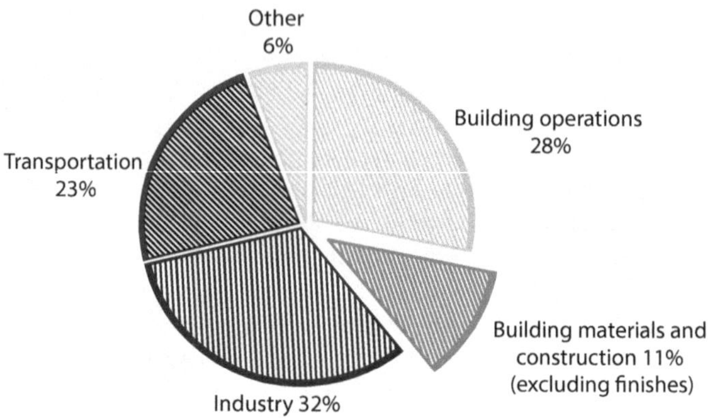

Plus, soaking up carbon dioxide from the air through forest plants' photosynthesis and turning it into solid materials like wood is a big way to cut down atmospheric carbon dioxide levels. That way, cutting back, swapping out, and locking up carbon altogether can better slow and curb climate change.

Buildings tie right into how folks live and work, so they're a hefty chunk of energy use and carbon emissions. Right now, cities big and small kick out about 70% of the world's carbon emissions, and buildings account for roughly 40% of the global total each year.

Like energy use, building carbon emissions are split into two big buckets: embodied carbon emissions and operational carbon emissions. Embodied carbon emissions come from making and hauling building materials and parts. Cutting embodied carbon means picking materials, structures, and building ways smartly at the design stage. Once a building's up, its embodied carbon's locked in—no changing it. Operational carbon emissions come from running lights, ventilation, heating, cooling—all the equipment a building uses. At the design stage, passive design can trim expected operational carbon emissions, and while it's running, upgrading systems and energy equipment can chop them down even more.

Let's circle back and weigh green building against low-carbon building. Both green and low-carbon buildings want to cut energy use, but green building goes beyond saving juice to look at biodiversity, human comfort, and cash efficiency too, while low-carbon building zeros in on saving energy, swapping fossil fuels for clean materials, and locking carbon with plant photosynthesis. These two ideas share a lot of ground but aren't quite the same.

You might notice some buildings—like the U.K.'s Beddington community we talked about—are both zero-energy and zero-carbon, showing that these ideas don't clash. Plenty of times, top-notch buildings pull off a "full-court press," nailing energy savings and carbon cuts together. But, bottom line, whether it's green or low-carbon building, the goal is human sustainability—a big principle all these ideas share, the root of it all.

Up to 2015, 82% of building energy still came from fossil fuels. You can see we've got a long way to go here.

04-2 What Does a Zero-Carbon Building Look Like?

Generally, a zero-carbon building means one with zero net carbon emissions each year. Like zero-energy buildings, a lot of what folks call zero-carbon buildings don't have zero carbon emissions over their whole lifecycle. Our research and practice on zero-carbon buildings are still in a feel-it-out, push-forward phase.

But don't sleep on this "operational zero-carbon"—pulling it off isn't easy; it takes real work. To go further for a true, full"zero carbon" life cycle, you've got to balance it out with all sorts of offsets.

Finished in June 2012 in Hong Kong, the"Zero Carbon Building"—ZCB—is one such zero-carbon demo building. Built by the Hong Kong Construction Industry Council and the Hong Kong SAR government, it's meant to show off low-carbon/zero-carbon tech, smart city ideas, and a low-carbon lifestyle with exhibits, education, and info for Hong Kong.

To craft a zero-carbon building, ZCB took three steps. Let's check out how they did it.

Step 1: Pick building materials and construction ways with low embodied carbon emissions. The ZCB went big with renewable, carbon-trapping materials like timber, bamboo, and plant fiber for floors, partitions, interiors, and more. For landscaping, they grabbed construction waste from the site teardown, boxed it in metal mesh, and used it for fencing.

Zero Carbon Building from above

Step 2: Use passive design and high-performance building systems to slash operational energy use as much as possible. The ZCB tapped Hong Kong's muggy climate with strategies like boosting natural ventilation, shading, roof insulation, and indirect sunlight for lighting.

Ventilation: The main building face points southeast, catching summer's main winds to cool things down. A corridor runs through the middle, setting up cross-ventilation to knock back the downsides of sticky, hot weather even more. Two wind catchers on the building's roof tap the wind pressure from rooftop airflow, pulling air down into the indoors to keep it cool.

Insulation: The south-facing roof is blanketed with a thick foam glass insulation layer, with 85% of the roof covered by solar panels and the leftover 15% a green roof.

Shading: The roof has overhang shades the outer walls and walkways, while vertical outdoor shading covers the glass walls on both sides.

Natural lighting: The northwest facade is the main light catcher, fitted with reflectors to bounce daylight deep inside. The roof smack in the building's center has light pipes too, making up for any dim spots

Summer prevailing wind direction

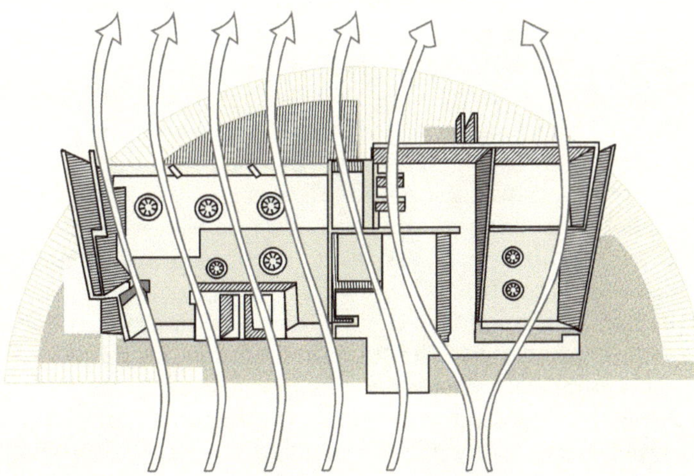

Front-to-back cross-ventilation

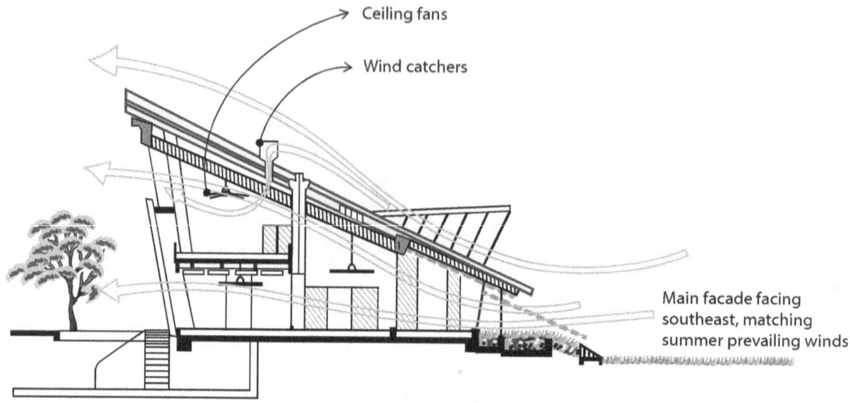

Natural ventilation section design

inside. The building's glass curtain wall system uses high-performance glass with low heat conductivity and high light pass-through, cutting heat radiation while keeping solid daylight. (This south-side shading, north-side lighting setup—doesn't it remind you of O'right Green Factory from last chapter?)

With this passive design, the ZCB nails the local muggy heat issue, slashing the run time of mechanical ventilation, AC, and indoor lights, saving 15%–20% more energy than standard-designed buildings like it.

On top of that, the ZCB rolled out a high-efficiency, energy-saving smart building system. For instance, big ceiling fans sit in all the public spaces, with cooling and dehumidifying systems split up—on slightly damp, stuffy days, just running the dehumidifier with fans gets it comfy. Plus, the building has an underground pre-cooling pipe system, using the temp gap between ground and surface to chill outside air before piping it in, easing the AC load. On top of that, it has adjustable skylights, heat-reflecting curtains, and other handy extras.

All that gear's run by a smart system, seriously boosting the building's overall performance. The smart building management system keeps tabs on ventilation, lighting, power, fire safety—auto-checking how it's running and tweaking the whole place's mechanical gear for top environmental performance. It has four microclimate stations too, sizing up how the building and its surroundings play off each other.

This whole "combo punch" setup shaved another 33% off the building's potential energy needs.

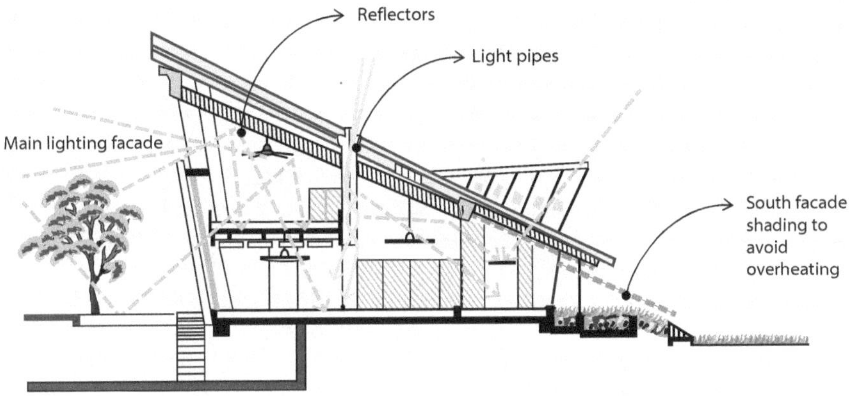

Natural lighting design

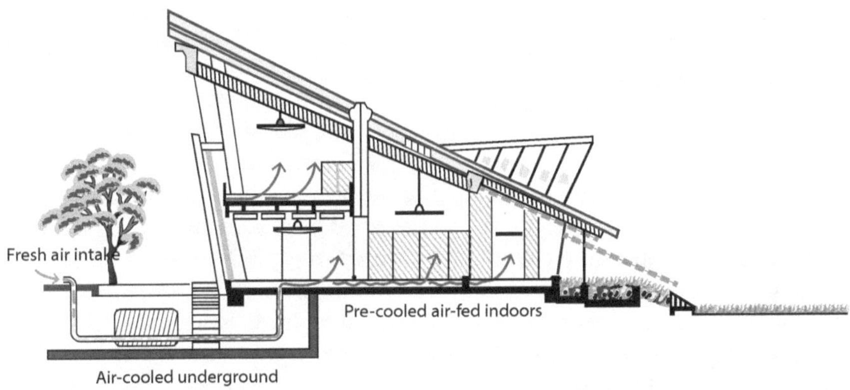

Underground pre-cooling pipe diagram

Step 3: Power the building steady and solid with renewable energy.

The Zero Carbon Building's big sloped roof is decked out with solar panels, pumping out 57% of its electricity. The building also has a bio-diesel trigeneration system, using biofuel from waste cooking oil to crank out power, with leftover heat from that juicing up cooling and dehumidifying.

The fact is, this project's renewable energy churns out more power than the building needs, with the extra feeding back into the city grid. So it's a straight-up "positive energy" building, no question. Its yearly power surplus hits about 15%, enough to slowly offset the embodied carbon emissions from materials and construction down the line, nailing true zero carbon for the full life cycle.

The ZCB's expected lifespan is 50 years. Estimates say over those 50 years, it'll offset about 7,100 tons of carbon dioxide—equal to saving over 11.5 million HKD (around 9.82 million RMB) in electric bills and the carbon-trapping work of more than 308,700 trees. Regular buildings burn energy and resources every day they run, but a sustainable spot like the ZCB chips in for the environment daily—the longer it's used, the better it gets for the planet.

Beyond that, the ZCB thought hard about protecting biodiversity and saving water too. It built a 2,000-square-meter urban native forest on-site with over 40 native tree types from Hong Kong, pitching in to protect wildlife, boost air quality, and tweak the microclimate. Plant growth can lock down more carbon dioxide from the air too, helping cut carbon. It has a rainwater collection system and water-saving irrigation setup too. Eco-toilets use low-flow fixtures and a graywater recycling system, saving 40% more water than standard equipment.

In 2021, the ZCB got an upgrade, adding a bunch of cutting-edge green building tech like a prefab construction showcase, self-cleaning and air-purifying solar glass canopies, low-upkeep plant walls, water-saving grass systems, aquaponic ponds, eco-friendly tree paths, and glow-in-the-dark jogging trails. Each update tweaks the building's energy use and environmental quality, nudging it closer to what folks dream of. Even in jam-packed, skyscraper-packed Hong Kong, where most buildings can't go this pure on energy and emissions cuts, the ZCB still shows off zero-carbon design ideas and lifestyles that fit the local climate, taking a big first step for Hong Kong's zero-carbon buildings. Plus, it gives this dense city a public spot for folks to chill and kids to play, carving out a corner in fast-paced urban life to stretch out and spark ideas. It often hosts creative markets, low-carbon weddings, green workshops, low-carbon concerts, and festival parties, pulling in more folks to get close to and grasp low-carbon living.

04-3 The Huge Potential of Old Building Updates

Since zero-carbon buildings have so many perks, shouldn't all new builds go zero-carbon? Sure, lots of countries are heading that way, but just hitting that mark isn't nearly enough.

Buildings outlast most everyday materials we use by a long shot. About two-thirds of the buildings standing now will still be around in 2050. The fact is that new builds are just a sliver of the total building stock. Compared to new builds, the heap of already-built buildings churns out way more energy use and carbon emissions.

Thanks to when they were built and old tech levels, lots of these old buildings didn't get high-efficiency designs—they're sucking down heaps of energy and pumping out tons of carbon every day they run. So we don't just need to slash carbon emissions full-on when planning and designing new buildings—we've got to revamp the ones already up, cutting the carbon they spit out while running too.

You Know What?

In China's green building push, upgrading existing buildings for energy savings is one of the top gigs.

By the end of 2015, China's northern heating zones wrapped up energy-saving upgrades on about 990 million square meters of existing homes, helping over 15 million households, saving 6.5 million tons of standard coal yearly, and seriously upping comfort in old digs. Plus, energy-saving upgrades for rural and public buildings are moving along step by step.

But at the same time, China's green building scene still has a bunch of hurdles to clear. About 60% of urban existing buildings are still energy hogs, and there's tons of room to spread renewable energy in the building game.

Speaking of old building revamps, there's one famous project you've got to check out—the German Reichstag redo in Berlin.

This century-weathered historic building, under British hotshot architect Norman Foster, didn't just get its grand mojo back—after the revamp, it slashed operational carbon emissions from about 7,000 tons a year down to around 440 tons. How does a built building chop carbon emissions that big with a makeover—what's the trick?

Built in 1894, the German Reichstag—originally the Imperial Diet—was a grand, solemn Baroque building with a central dome. After taking hits from fire and war in World War II, the central dome was gone, and the insides were shot full of holes. Post–Cold War, Germany's parliament decided to reclaim the Imperial Diet as their office spot, renaming it the Reichstag.

How do you fix up a building this big deal? No doubt, it's a tough, tangled job. After a contest with 80 German architects and 13 foreign

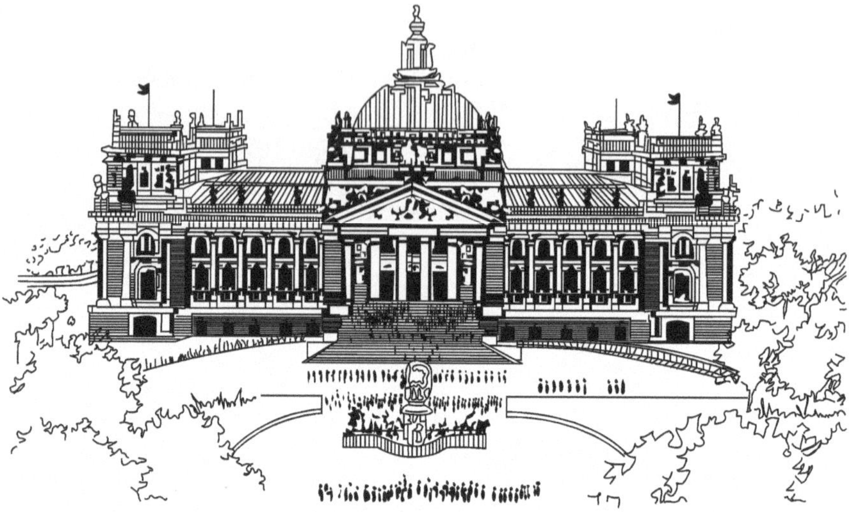

Original Imperial Diet look

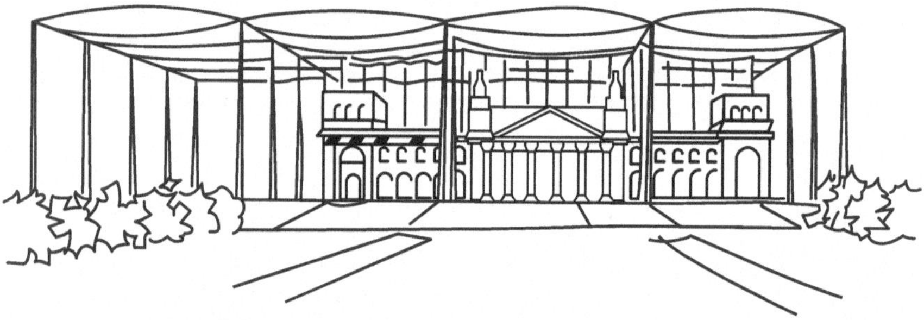

Norman Foster's first draft sketch

firms, three plans made the shortlist. One was Norman Foster's pitch to cap the old building with a see-through glass roof.

In round 2, cash realities forced the three finalists to start trimming their makeover budgets. Foster went back to square one, tossing out four backup plans. Since the old Imperial Diet's classic shape was so etched in folks' minds—and as the Reichstag, the dome carried heavy political vibes—Foster had to bend, designing a glass dome for the new setup. Even though the inside's functions and layout shifted from the past, the outer face and dome shape stuck around, holding onto the

German Reichstag (after restoration)

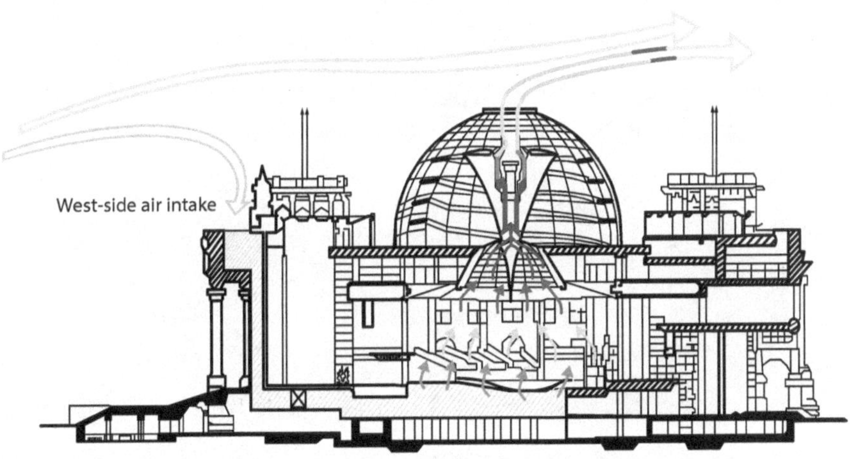

Passive ventilation design

city folks' memories of their twisty history and the building's pull as a landmark.

Don't let the dome fool you—Foster's version and the 19th-century one aren't the same beast. Truth is, that dome—super controversial at first—ended up the most eye-catching part of the whole redo. It has a 40-meter diameter and stands 23.5 meters tall. Foster smartly mixed in passive design, turning the glass dome into a primo way to pull in natural light and whip up natural ventilation.

For ventilation, fresh air gets pulled in from the west-side intake, seeping slowly into the parliament hall through tons of tiny vents in the floor. Like that chimney effect we talked about, heated air rises to the hall's top and then gets sucked out through ducts in the central upside-down cone under the dome.

For lighting, the dome's sunshades block some of the harsh glare, while the rest of the light bounces off 360 mirrored glass panels in the middle cone, landing in the parliament hall below. Plus, the architect threw in a 240-meter spiral ramp inside the glass dome, luring visitors up to the observation deck. Up there, folks can look down at the Reichstag's shimmering glass dome and out at Berlin's exceptional city views—a must-hit "check-in" spot for tourists.

Dome interior

Meeting hall below the dome

Hit the dome's tip-top, and the parliament hall's insides are all laid out below—hands down a bold breakthrough in the redo design. It wipes out that cold, stiff, dull vibe of political buildings, handing residents a sleek, high-tech, future-feeling space that's way more approachable.

Maxing out natural light and ventilation already cut plenty of energy and carbon emissions, but the architect didn't call it quits there. The old Imperial Diet's built-in power system ran on mineral fuel, spitting out up to 7,000 tons of carbon dioxide a year. The new Reichstag's combined heat and power setup uses biofuel pulled from plant oil—a renewable, clean fuel that slashes carbon dioxide emissions big time. Plus, the building's roof has over 300 square meters of solar panels. These moves cover more than 80% of the new Reichstag's and nearby parliament buildings' electrical needs.

Berlin's climate swings cold in winter and hot in summer—heating and cooling eat up a huge chunk of the building's operational energy. For that, the Reichstag tapped a local-fit underground heat storage system, stashing cold water about 60 meters down in a chilly aquifer for summer cooling, and banking extra heat from summer combined heat and power about 320 meters deep in a warm aquifer for winter heating. Researchers' monitoring and math show this system runs slick and steady, losing no more than 30% of stored heat on the way.

Blending a jaw-dropping glass dome with a bunch of passive design strategies and high-efficiency systems, Norman Foster and his crew made compromises and tweaks facing real-world snags but still breathed fresh life into this nearly wrecked old building.

Truth is, architects often play that role—cooking up their own ideas and innovations while respecting and juggling what clients, builders, users, and other stakeholders want and need, checking if the cash holds up, and teaming up with structural, electrical, and HVAC engineers. It's only when all those stakeholders swap and clash ideas that a wild "chemical reaction" sparks—often landing you with surprises you never saw coming.

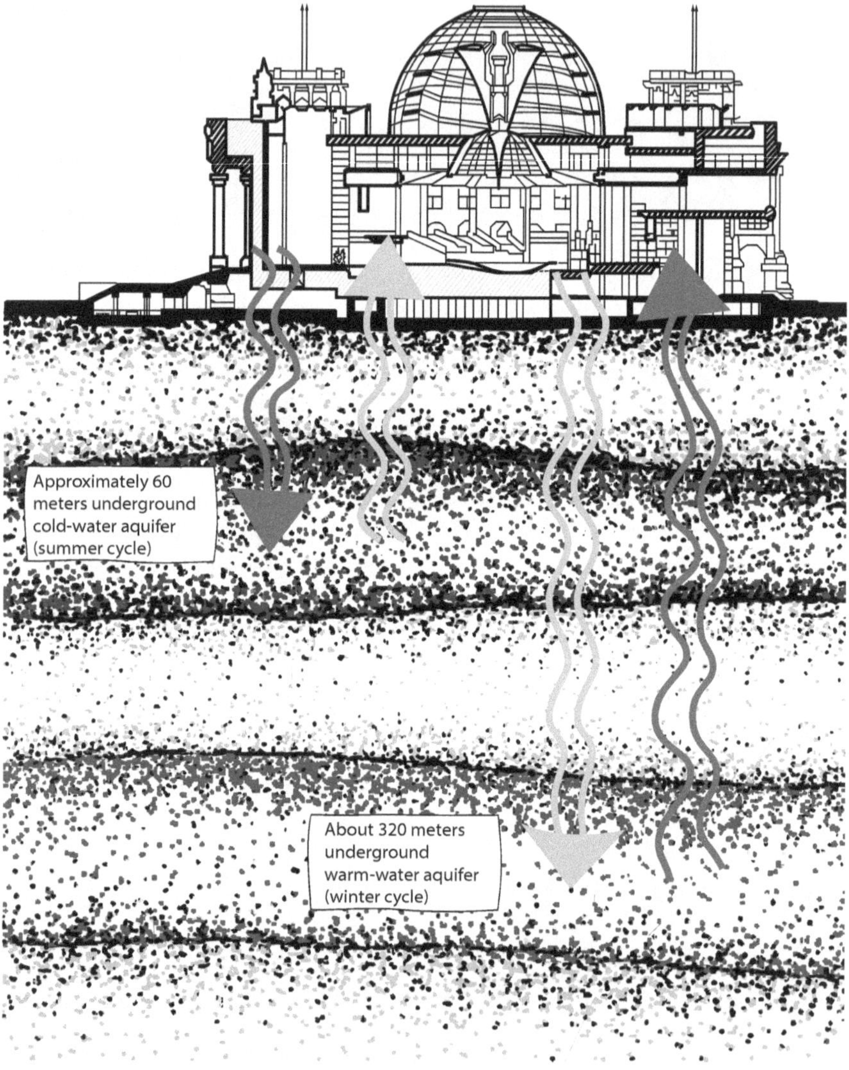

Underground heat storage system diagram

You Know What?

Not all existing building revamps are as "fancy-pants" as the Berlin Reichstag. The fact is that in China's cold zones and hot summer–cold winter spots, home energy-saving upgrades have been chugging along steadily, helping tens of millions of folks. According to the 2012 "Guide to Energy-Saving Retrofits for Existing Homes" from China's Ministry of Housing and Urban-Rural Development, home energy upgrades boil down to four main chunks:

1. Insulation upgrades for outer walls, roofs, doors, windows—all the wrapping materials
2. Tweaks to heating systems for per-house metering and room-by-room temperature control
3. Energy-saving fixes for heat sources—like boiler rooms or stations—and heating networks
4. Full-on energy-saving upgrades covering building repairs, function boosts, and renewable energy adoption

Unlike new builds, old building revamps usually happen with folks still living inside—you've got to dig deep into residents' situations, nail the public outreach and chitchat, clear up questions, get their okay and teamwork, and keep both them and the workers safe. After the revamp's done, you've got to hand out user guides too, showing residents how to use the energy-saving designs right and build solid energy-saving habits.

05 Cooling Down Cities

Buildings don't stand alone—they team up with nearby roads, plazas, parks, and all those man-made materials to make a whole. We need to think big and systematically about sustainable building environments—like on a city scale. Beyond nailing down scientific, detailed building designs, what else in city-building ties into sustainability? Let's kick off by seeing how to cool cities down.

05-1 "Feverish" Cities

Not sure if you've felt it, but summers in cities seem to be getting hotter. These days, AC's old hat—we feel like summer's unbearable without it—but the truth is, it only become common in the last few decades. Back then, not every home had AC—plenty of offices and public spots didn't either—but folks got by just fine. So why can't we hack summer without AC now? Have we gone soft?

You might be thinking back to that global climate change we hit in the last chapter. Sure, climate change cranks up wild weather and adds more scorching summer days, but that's not the only reason cities are heating up. For one, our tolerance for indoor temps might've dropped. Back then, a little heat with a fan was no big deal—now we reckon crisp and sweat-free is the real cool. On the flip side, our cities are flat-out getting hotter. The urban heat island effect is a big driver jacking up temps in these "concrete jungles."

What studies show now is that, as cities build up and grow, the urban heat island effect keeps getting worse year by year. The heat island effect's downsides—like less comfy outdoor spaces, more building cooling energy, and extra sick folks in hot spells—are grabbing more attention too.

You Know What?

The urban heat island effect is when city temps run hotter than rural spots. Way back in the early 20th century, British climate folks floated this idea. After the 1960s, people started snapping infrared pics of Earth with satellites—those shots made urban heat islands easy to spot, proving the theory.

The urban heat island effect comes down to five main reasons:

1. Less greenery in cities cuts down on ground evaporation cooling.
2. Materials like concrete and asphalt—building materials with higher heat capacity and conductivity—soak up and stash more solar heat.
3. Solar heat bouncing around tight-packed buildings heats up the city environment.
4. City buildings block wind flow, clogging up the city's heat escape.
5. City cars, industrial equipment, home appliances, and people themselves pump out heat, pushing temps higher.

Urban heat island effect

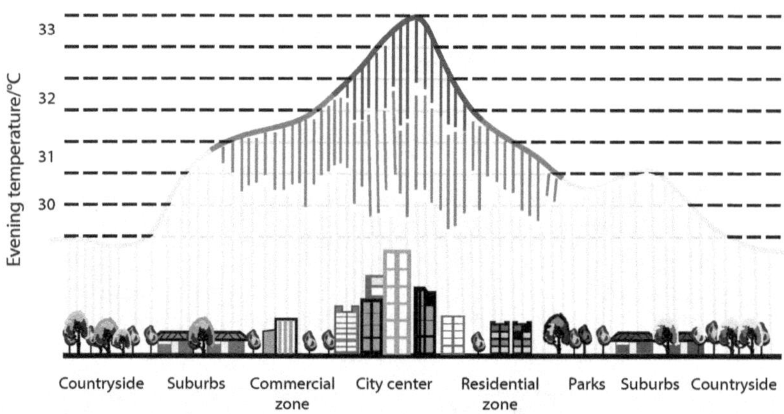

The heat island effect piling onto the greenhouse effect sounds heavy, but we're not totally out of moves. City planners have cooked up plenty of ways to dial back the urban heat island effect:

Bumping up city greenery and water boosts evaporation cooling, adds shade, cuts heat soak-up, and eases the heat island effect.

Buildings with high-reflectivity light-colored outer surfaces, plus a solid insulation layer, can cut down on soaking up and holding solar heat.

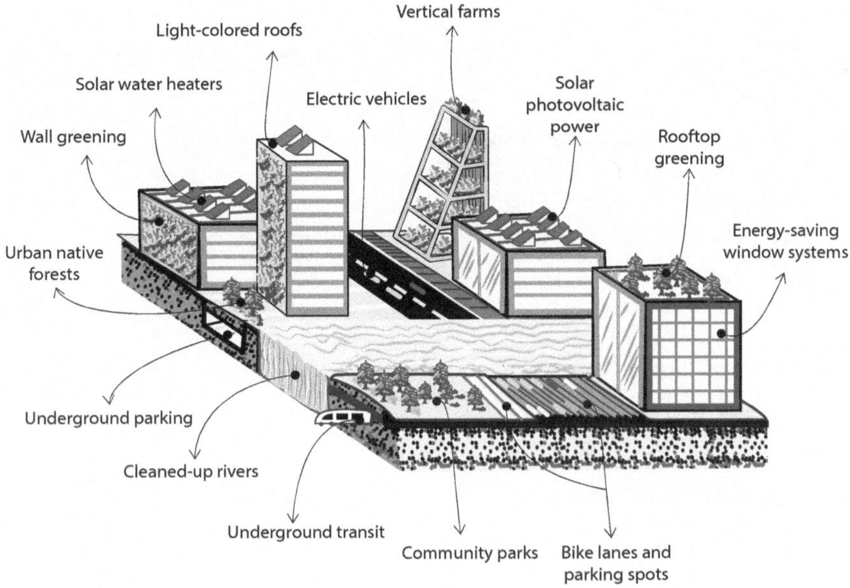

Green/livable cities

Setting up city wind paths right, leaving enough ventilation openings in planning and building design, boosts airflow and helps shed heat.

Pushing walking, biking, and public transit over driving cuts down on the cars on the road and dials back AC and other energy-guzzling, heat-spitting equipment, trimming city heat sources and lending a hand.

You Know What?

In muggy, hot "concrete jungle" Hong Kong, developers once threw up tons of high-rise row homes on seaside lots to cash in on "ocean-view flats." They stood like a screen, hogging the best views but choking off the sea–land breezes crucial for city airflow. Local media's rough count pegs Hong Kong at 104 of these "screen buildings," potentially hitting 1.05 million folks.

In 2003, SARS slammed Hong Kong. The Hong Kong SAR government clocked how big a deal city ventilation is too. The government tapped Professor Wu Enrong's team at the Chinese University of Hong Kong's

School of Architecture for years of research, rolling out the Air Ventilation Assessment System—AVA—in 2006, putting screen buildings in the past for good.

In 2016, China's National Development and Reform Commission and Housing Ministry dropped the "Urban Climate Change Adaptation Action Plan," calling for city planning to "open up ventilation corridors, boost air flow, and ease heat islands and smog." Hong Kong's research and know-how turned into a key playbook for mainland cities too.

05-2 A Hotel Growing a Tropical Rainforest

In "summer-all-year" tropical rainforest climate Singapore, greening us a big move to cool off urban heat islands and lift the city's vibe. Building a "garden city" is a major thrust in Singapore's city planning.

Singapore's WOHA firm, all about mixing sustainable design with local settings, keeps trying to weave that "garden city" idea into its projects. Finished in 2013, the Park Royal on Pickering in Singapore is one of the firm's standout gigs.

This hotel sits smack in the middle of Singapore's downtown, right by Hong Lin Park, the Central Business District, Chinatown, and Clarke Quay. In this bustling, tight-packed city core, just keeping ground-level greenery wasn't enough to match the architects' "big dreams."

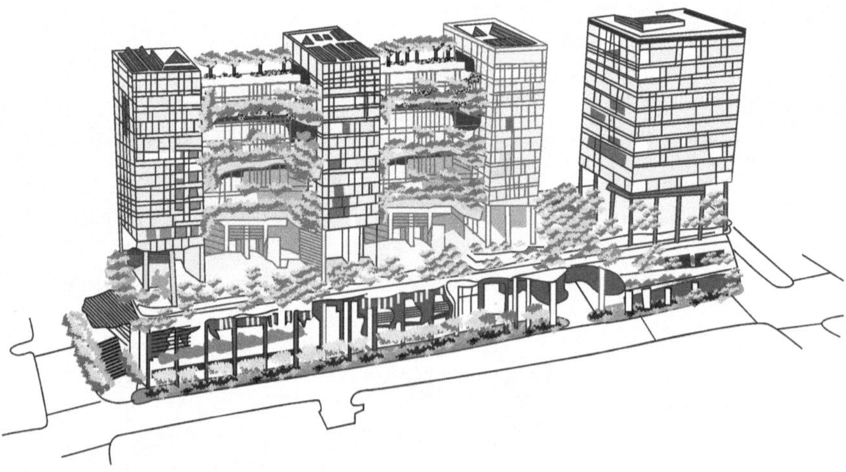

Park Royal Hotel

They wanted to stack more green space across the building's floors, giving back extra scenery and environmental perks to the city around it. This approach jumps past the usual "minimalize bad materials" thinking, straight to "keep dishing out good materials"—no doubt a fresh twist on green building.

The ground floor rocks an over-the-top raised design with cave-like layered shapes, blending with street-side plants to whip up a half-open public space along the city front. In some spots, the ground level is fully lifted, linking the front and back blocks straight through. This give-the-ground-back-to-the-city design shows the architects' big-picture take on downtown public spaces, carving out ventilation paths, spotlighting local flavor, and making you want to stare and dig deeper.

Up on the fifth-floor pool deck, water features and green plants play off each other very nicely. Every four floors up, jutting sky gardens pop out, building a 3D greenery setup—like a tropical rainforest tucked in the concrete jungle.

Those sky gardens don't just flash lively green views for the rooms—they block the blazing sun from hitting straight on too. Even with big glass walls in the guest rooms, you don't have to sweat overheating. This angle faces right toward Hong Lin Park, letting you soak up the park's views with nothing in the way.

The hotel's sky gardens stretch over 15,000 square meters—double the building's footprint and matching the size of nearby Hong Lin Park. The plant mix is all over the place—tall palms, shady broadleaf trees, low shrubs, blooming flowers, climbing vines, you name it. This wild variety of plants doesn't just give folks shade and pretty views—it's a hangout for city critters too, pitching in for urban biodiversity and ecological balance.

Are you wondering if planting all these greens makes watering and upkeep a hassle? If that crossed your mind, you've already got the hang of thinking in full life-cycle terms.

Sure thing, finishing the building is just the start—keeping all these plants alive isn't a cakewalk. The Park Royal's plants get watered with rainwater, and when that's not enough, recycled, cleaned-up non-drinking water fills the gap. Rooftop solar panels power the irrigation water cycle and lighting systems.

Plus, most plants here are local species—best suited to the climate, they thrive without much human fussing. That cuts down on upkeep costs and headaches too. Of course, these fixes go beyond just architects—you need pros in structural design, clean energy tech, water treatment, plant growing, and more, all teaming up across departments to nail the right answers.

Hotel entrance space

Even with its big greenery system, the Park Royal didn't trade off performance—it actually dropped energy use and boosted efficiency. This building was awarded the platinum Green Mark from Singapore's green building rating system—the national, top-tier green building badge. (What's this green building rating system and certification? I dig into that in Chapter 8.)

Sure, every strategy has its ups and downs—you must use it in the right spot to max it out, and building greenery is no different. While plants help ease urban heat islands, some studies say in jam-packed cities, too-thick greenery might clog up ventilation. That means when designing cities and landscapes, you've got to weigh plant spots, heights, and even leaf thickness and seasonal shifts all together.

Sky gardens

You Know What?

Planting in cities comes with these perks:

1. Blocking solar radiation, keeping ground and building surfaces from soaking up too much heat
2. Cooling ground and building surfaces by evaporating water
3. Giving city critters and bugs a place to crash
4. Sucking up some of the air's gunk, cleaning up air quality
5. Soaking up carbon dioxide, slowing the greenhouse effect
6. Sprucing up the scene, giving folks' eyes and souls a breather
7. Planting some fruits and veggies right can stock city folks with fresh eats

Plants are alive—growing and tending them takes people, materials, energy, and water. So picking plant types, designing upkeep systems, and sourcing irrigation water—there's a lot to get right. Like in spots with harsh climates or lousy light, you need plants that can handle drought, cold, or shade, depending on what's real. In places with summer typhoons and downpours, you've got to watch out with roof and vertical greening—if the support structures get trashed by wild weather, it could turn dicey.

In 2016, City University of Hong Kong had one of those accidents. On May 20 afternoon, after days of rain, City University's gym roof caved in—metal framing, concrete, and rooftop grass crashed inside, freaking out the students and staff there. That collapse hit 1,400 square meters, injuring three people. Good thing few students were in the gym then, or it could've been a nightmare.

In a dense, hot-summer city like Hong Kong, roof greening seemed like a solid design move. But the accident probe showed the gym's roof greening used wrong data and info in design; plus, it had drainage hiccups. After steady rain, too much water piled up in the soil, overloading the roof beyond what it could take, sparking a major disruption.

In summary, any green building strategy must be placed in an appropriate location and designed with scientific precision to fulfill its purpose effectively. If green building has a rule, it's that there's no one-size-fits-all, set-it-and-forget-it fix. Science and local know-how always go hand in hand—you can't skip either.

You Know What?

In jam-packed, land-scarce cities and hilly spots with uneven ground, architects flipped greening from flat to upright. Compared to old-school ground planting, vertical greening saves space and fits all kinds of terrain. It doesn't just cut down solar heat soaking into or bouncing off building surfaces, dodging too much heat piling up in city canyons—it also eases that boxed-in feeling dense cities dish out.

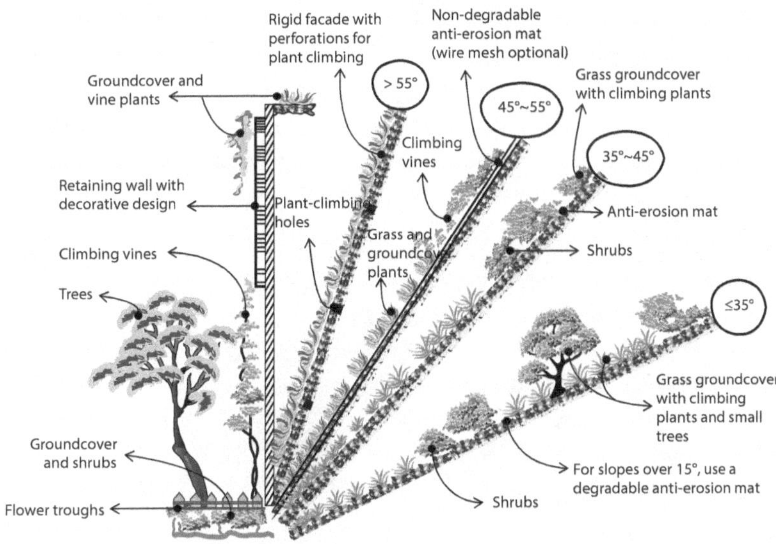

05-3 A Cool, Dreamy Urban Wharf

Cooling a city is a big-system job—beyond chilling high-rises, you've got to cool public spaces too. After all, folks can't always hide inside soaking up the cool AC—you must step out, move around, and mingle outdoors to keep city dwellers' bodies and minds in good shape.

Cooling outdoor spaces is no small trick. Outdoor spaces stretch every which way with people buzzing around—how do architects wrangle the temperatures? The 2006 Clarke Quay redo in Singapore took a gutsy swing at it.

Clarke Quay has deep roots—used to be a trading hub famous for its flea markets. But the old wharf was clogged with squat buildings,

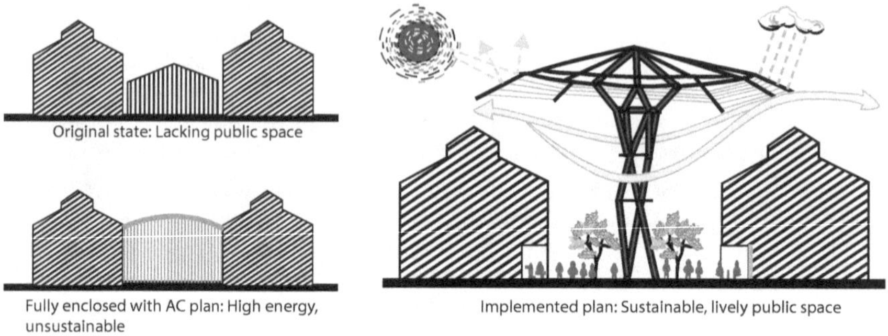

Clarke Quay revamp strategies

dull streets, and nothing special. With the local heat and rainy season downpours, the wharf's shopping streets are often draped in makeshift canvas tents. Even if it helped some with the sun and rain, the low, messy setup didn't pull many tourists.

To spark business growth and breathe life back into the rundown, chaotic wharf, developers kicked off a revamp in 2002. This redo didn't just tidy up the wharf's shops, bring in top-notch stores, and lift service quality—it gave public spaces a major glow-up too. The project was handled by SPARK Architects.

Singapore has year-round warmth, with long summers—a sticky, hot tropical rainforest climate. Leaving streets wide open brings overheating headaches, but sealing them all up isn't very practical either. The design crew got crafty, capping the wharf's walking streets with a shading system. Giant umbrella-like shades sprout up from the middle of the walking streets, cutting solar heat and fixing the rainy season umbrella hassle. The umbrella setup arches up, rising above the buildings on both sides so breezes can slip through the middle. These big umbrella wings stretch over the rooftops on both sides, cooling the walking street while shading the buildings too, balancing indoor and outdoor comfort.

Just shading isn't enough. The design crew tossed in dry fountains called "water curtains" at the street ends and insides too. As convective air passes over the water curtains, it is naturally cooled. Due to the "chimney effect," the umbrella-like structure's top heats up under the sun, creating a passive updraft that pulls cooler air from below. To back that up, some umbrella parts have active exhaust systems with fans at the top, steadily pushing indoor air upward. This way, air cooled by the water curtains flows into the walking street, exits through the umbrella tops, carries off extra heat, and sets up a steady cool-breeze cycle. Folks

Clarke Quay revamp project from above

"Water curtains"

on the walking street can feel a gentle breeze, easing the sticky, muggy discomfort.

Research data shows that the Clarke Quay redo can drop walking street temps by up to 5°C in the blazing summer heat. Beyond sprucing up the wharf's commercial microclimate, this set of giant umbrella structures nabbed tourists' eyes, turning into a new Singapore landmark and a must-hit spot, drawing over 2 million visitors yearly. The owners' profits shot up five times over what they were before the revamp. Clearly, caring about visual appeal and boosting environmental performance don't clash. With smart design, they can lift each other up and score multiple wins. If you ever get to Singapore, don't miss checking this place out.

You Know What?

As people and resources keep piling into big cities, rows of skyscrapers seem to stand as the go-to sign of a thriving, bustling place. For these packed cities, setting up solid **city ventilation** is a key move to ease the urban heat island effect. City ventilation isn't just about single building designs—it has to factor in bigger scales like neighborhood, city, and urban planning too. Here are some ways to better city airflow:

Ventilation corridors: City planning should mesh with local prevailing winds, carving out airflow paths through streets and public spaces to pull breezes into every corner.

Building coverage and permeability: In cities, the wind at the pedestrian level hits our daily lives the hardest. So the podium parts at a building's base shouldn't be too packed or solid. You can raise the ground level, break up podium chunks, or step back floors to leave more room for pedestrian-height breezes. In super-dense zones, you need ventilation openings mid-to-high-up too, to lift airflow in those building spaces.

Building layout and height: How buildings face and line up ties tight to ventilation. Like leaving enough space between buildings, aligning layouts with prevailing winds, or staggering front and back rows—all help spruce up the wind scene. In spots like coasts or hills with clear prevailing winds, building heights should roll with the breeze—low upwind, high downwind—not sticking the tallest buildings right in front.

Of course, these city ventilation strategies have their limits on where they work. In real cold climates or spots with wild wind setups, the core wind issues cities a face shift. So, what's the deal with the city you live in?

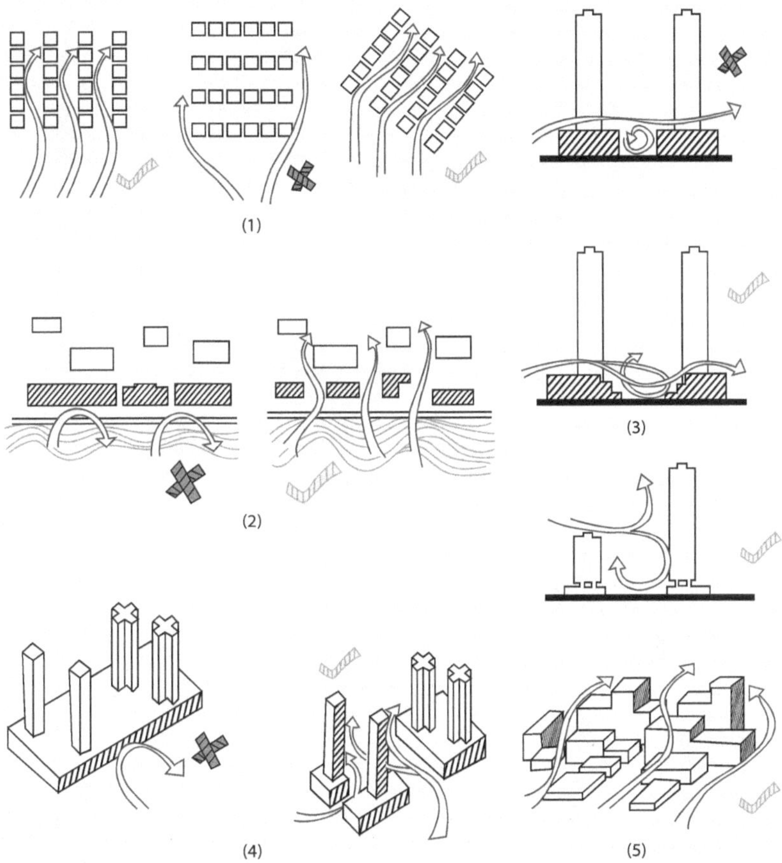

City ventilation

06 Being a Smart User

When you step into an office building, the overhead lights flick on with your footsteps and switch off a bit after you leave. The office windows' glass turns frosted bit by bit as sunlight ramps up. Solar panels track the sun like sunflowers. The AC kicks on or shuts off on its own based on indoor temperature shifts. With automated, smart building systems spreading, these once "sci-fi" scenes are slipping into daily life—sometimes we don't even notice them.

But can machines and computer systems really make the smartest calls for us? Does a smart building mean swapping human decisions for slick equipment, or using tech to max out human ingenuity? Have you ever given this question some thought?

06-1 Precision-Running Building Machines

Universities worldwide have long been key hubs for scientific research. In architecture, many top universities nurture research centers and teams focused on intelligent buildings. Carnegie Mellon University in the U.S. hosts the Center for Building Performance and Diagnostics (CBPD), studying green building comfort, energy and environmental benefits, and tech applications to ensure these buildings, once built, deliver expected energy savings, environmental gains, and comfort during use. For buildings that don't perform well in practice, the CBPD acts like a doctor, offering "diagnoses" and fixes.

The CBPD's office—the Intelligent Workplace—is itself a stand-out smart office building. It "perches" on the fourth-floor roof of an old campus building—Margaret Morrison Carnegie Hall—just one story, about 600 square meters. As a rooftop add-on to an old building, it keeps a low profile and looks plain at first glance; built in 1997, it's not exactly new either. But none of that stops it from being a model smart

Intelligent Workplace

building today—it's a "living" experimental space. From the start, the design team set it up to keep updating by loading in cutting-edge systems, components, and materials. It's like a house with a pulse—metabolizing, growing, evolving nonstop.

Right from the design kickoff, the team used advanced building info computer simulation tech, tapping professional software to model and sort every building part, material, and system in detail. This way, the team could pinpoint every building piece's material, count, spot, and performance in the software, swapping or picking materials later as needed. With this hefty, spot-on computer building information model, the team could plug it into energy and performance simulation software, testing how different designs play out to find the one that best fits the local climate, saves energy, and keeps things comfy.

Take an example. The Intelligent Workplace's lopsided sloped roof design came straight from computer-simulated optimization. Based on local sun paths, this roof slant catches maximum daylight, cutting down on artificial lighting energy. But more light isn't always better—skylights that are too big can overheat rooms and throw glare. So the team ran computer sims comparing two roof sizes and window setups—A and B—landing on B as the more energy-saving, comfort-balanced pick.

Like those other green building cases we've looked at, the Intelligent Workplace packs tons of passive design, layered with a bunch of high-efficiency building systems.

Computer simulation tech aiding roof design

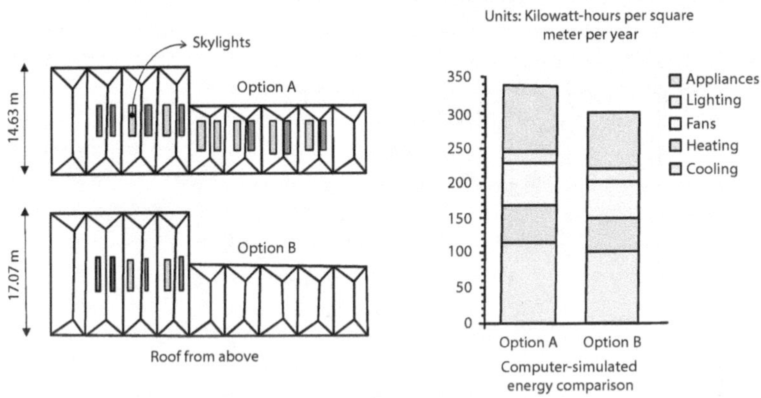

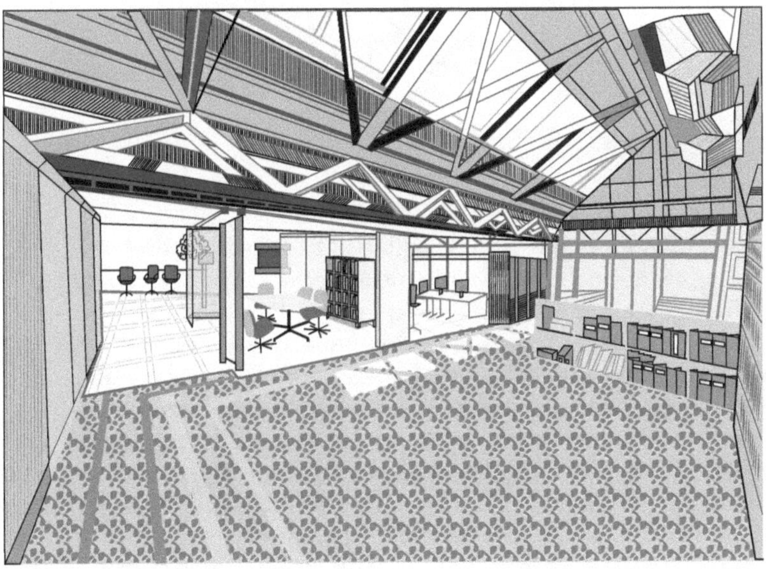

Intelligent Workplace interior effect

During milder chunks of the year, natural ventilation through windows meets the need—more energy-saving than AC and is better for users' health, body, and mind. That's why window design is a big deal in green buildings.

The Intelligent Workplace's windows use double-layer hollow glass to enhance insulation and heat retention. The glass surface features a coating that reduces infrared light transmission, minimizing excess heat

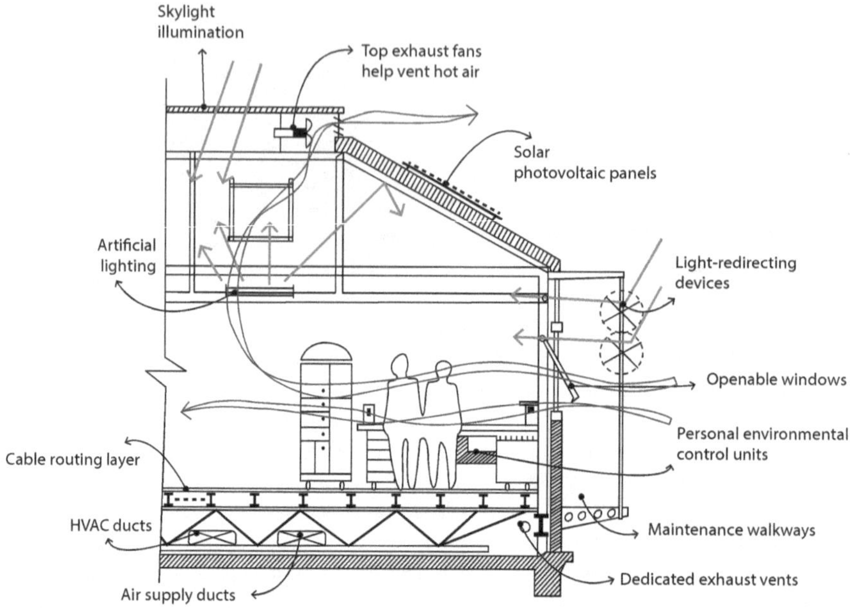

Intelligent Workplace design strategy diagram

radiation. At the same time, 70% of visible light passes through the glass, delivering a bright and comfortable indoor effect. The seemingly plain window frames hide a clever twist—they're fitted with a hot-and-cold water circulation system. Cold water circulates in summer and hot water in winter, all to make folks standing by the window feel more at ease.

The Intelligent Workplace puts serious effort into harnessing natural light. The building's outer walls have light-redirecting devices that adjust the incoming light direction based on the sun's path and weather. By mechanically tweaking reflector angles, these devices channel maximum natural light indoors while cutting glare. A gap between the reflectors and outer wall lets maintenance workers pass through and makes opening wall windows a breeze.

The building's indoor lights don't beam straight down—instead, they shine up at the ceiling. Light scatters off the ceiling, spreading evenly indoors, making the artificial lighting softer and more natural.

This blend and balance of passive design with high-performance gear drops the building's energy use to a minimum while maxing out comfort. What stands out is the building's heavy use of remotely operated mechanical systems, enabling finer-tuned smart control from afar.

You Know What?

Poorly designed buildings can also make people sick.

As far back as 1983, the World Health Organization coined the term sick building syndrome. Researchers found that in air-conditioned office buildings, lots of workers showed up with asthma, allergies, coughing, sneezing, dizziness, and drowsiness all at once. But when workers left the building or took time off, those symptoms eased up. That's what we call "air-conditioning sickness" in everyday terms.

Sick building syndrome isn't a strict medical condition—it's more about showing how crummy building environments can hit people's physical and mental health in a big, messy way. Bad ventilation, chemical pollution, biological gunk (like mold in ducts), and off-kilter temps, humidity, and lighting can all spark sick building syndrome.

A 1997 study says in the United States, fixing building environments could cut $17 billion to $33 billion yearly in costs for treating respiratory issues, asthma, and sick building syndrome while saving another $12 billion to $125 billion tied to boosting worker performance.

Sure, upgrading HVAC, lighting, ventilation, and air quality in buildings takes a hefty chunk of manpower, materials, and time. But researchers reckon if you look long-term, investing in better building environments pays back 18 to 47 times over.

Sure, these estimates vary in the nitty-gritty, and there's plenty to hash out on the practical side, but tons of studies point to one truth: Building environment quality is tied tight to human productivity and well-being. Boosting building environment quality can dish out massive social benefits.

If you were the company boss, how'd you decide?

The Intelligent Workplace's "ingenuity" not only shines in the design phase.

While running, sensors all over inside and out keep track of the building's and surrounding area's temperature, humidity, airflow, sunlight strength, rainfall—you name it—logging it all into a computer data system built just for the Intelligent Workplace. Users can peek at all that data anytime on a networked computer or phone and tweak the building's gear to their liking based on the data. For example, if the outdoor temp and humidity look cozy that day, users can switch off their seat's AC vent and crack a window.

What's great is, even when not in the office, users can hop online with a computer or phone to check the building data system and run equipment from afar. For example, when a user gets home and sees the desk lamp above their workspace still on, they can switch it off via their phone. If they're away on a long trip, users can check their office anytime and cut power to their spot's outlets through their phone, saving electricity.

Beyond real-time data, users can also look at historical records, reviewing and assessing building energy use and patterns over a past stretch to spot what's working well and what needs a tweak.

This system fully taps users' initiative, letting them judge and adjust their actions anytime to use the building better. If you had this system, would you find it a hassle, or would you dive in with curiosity and an open mind to explore it?

The Intelligent Workplace has another special perk—its adaptability and upgradability.

To test new green building tech and strategies, researchers need to swap out, tweak, run, and remove some building systems from time to time. That's why the building's insides use four "plug-and-play" module systems.

First, the building's HVAC, lighting, power, comms, and computer support items aren't fixed to walls—it's tucked overhead and under floors, smartly designed as plug-and-play modules. These modules can shift spots as needed or pop off anytime for checks and adjustments.

Second, the building's furniture, partitions, and lightweight dividers aren't locked down—they can move whenever needed.

Third, the building's projectors, sound systems, multimedia equipment, and computers are mobile and plug-and-play too.

Fourth, the building's energy supply and delivery systems—like outer wall modules, cooling modules, biomass energy modules—aren't set in stone; they're bolted together for quick swaps, flexible repositioning, and plug-and-play use. The building's outer facade has maintenance walkways too, making it easy to swap or fix wall parts.

This approach ramps up the building's flexibility to the max, cutting down big demolition and installation jobs during space revamps and equipment upgrades.

Through this tight, system-wide planning, plus teamwork from architects, engineers, and pros across fields, the Intelligent Workplace turned into this one-of-a-kind smart building. Even though it was built over 20 years ago, its solid indoor environment quality and top-notch energy savings prove it has delivered on its design goals to the fullest. Compared to U.S. standard building energy use per square foot, the

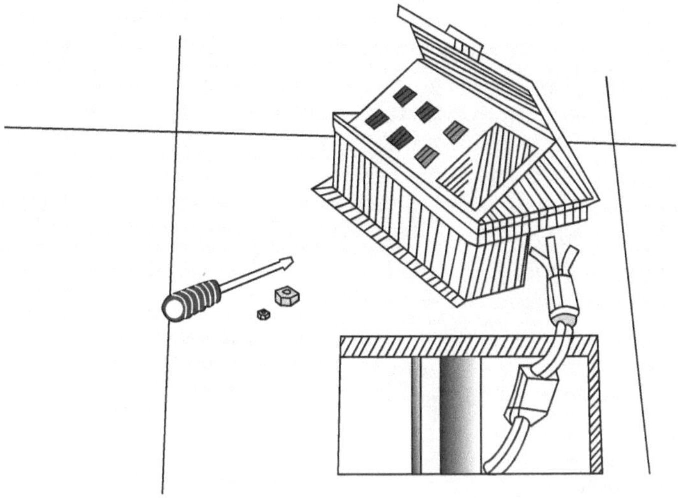

Movable outlets and switch units set in the floor

Intelligent Workplace uses just one-fifth of the energy to hit stellar air quality, thermal comfort, lighting comfort, and indoor space effects.

You might say the Intelligent Workplace's slick "evolution" chops come from its research building status—regular folks' buildings can't hit that peak—and I'd agree. But its big-picture, system-thinking approach; cross-field teamwork; forward-looking logic leaving room for change and growth, and human-centered push to tap users' initiative are all worth borrowing and mulling over.

These days, our lifestyles keep diversifying, standards aren't one-size-fits-all anymore, and tech is changing fast. Meanwhile, buildings' lifespans stretch 70 years, even centuries. What kind of building stands the test of time and nails true "sustainability"? There's no textbook answer for that. If it pops into your head now and then, poking at it from different angles can be pretty engaging!

06-2 Are Users a Disruption or a Bonus?

The first half of this book spent a good chunk digging into how buildings should be designed. So once a building's up, does it always hit the mark set at design time?

We know that in tons of green building designs, engineers start by simulating energy use and performance with software. Usually, those sims need a bunch of "assumptions"—climate data, user basics, daily use hours, surrounding terrain, and so on. Those assumption numbers are typically theoretical, idealized figures, so the results come out theoretical and idealized too.

But in real use, things often aren't that picture-perfect. Climate change may result in anomalous weather patterns, significantly deviating from historical climate data. A new high-rise might be constructed beside the building, obstructing sunlight. Human activity is even harder to predict—some frequently forget to turn off computers, while others keep lights on during broad daylight. Another obvious yet difficult-to-avoid issue is that contractors, to cut costs or save effort, may not strictly adhere to the design specifications during construction. Thus, the theoretical energy consumption values from the design phase and the actual energy consumption values during operation can differ vastly. No wonder some engineers jokingly say that people are the biggest "disruption factor" in building energy-saving models.

So here's the question: after all this complex computer simulation, design, and tuning, once the building's up and running, should we minimize human "interference" and let smart systems take full control to ensure the building operates at its best?

Let's set that question aside for now—I'll tell you a little story.

It was a sweltering summer while I was pursuing my PhD in Hong Kong—we were in a classroom with blasting AC, most of us in short sleeves. The room was downright frigid—we tried nudging the AC temp up a bit, but after a few attempts, we found the system was locked; no manual adjustments were allowed. During the break, fed up, we reached out to the building manager, hoping he'd unlock it and bump the AC temp up a touch. To our surprise, the manager told us that for energy saving and eco reasons, school rules banned users from tweaking classroom AC temps at will.

Clearly, the classroom temp was set by the building folks at a "theoretically" comfy, energy-saving number. The managers wanted to stop users from cranking the AC too low and wasting energy needlessly. But sadly, this temp that they thought was ideal might suit suit-and-tie office workers, but for us in short sleeves, it was way too chilly. What's ironic is that day, we were taking a green building course. After all that fuss, the class got a real hands-on lesson.

You Know What?

Everyone's take on thermal comfort and needs varies—it's tied to metabolism rates (usually linked to age, gender, ethnicity, etc.), clothing thickness, and things like air speed and humidity in the environment.

Folks notice female office workers often gripe the AC's too cold—they've got to keep an extra shawl or jacket handy. The AC system's clearly "smart" enough to figure out a comfy temp range for us—so why's this still an issue?

Researchers dug deep and finally pinned down the culprit. Turns out, lots of AC systems lean on thermal comfort models cooked up back in the 1960s. That model's metabolism rate numbers came from a guy about 40 years old, weighing around 70 kilos. Maybe back then, the research crew figured this kind of guy was the most common office type, the best stand-in. But nowadays, women are a big segment of the office crowd, and their builds are usually smaller than men's, with more body fat and less muscle, so their metabolism rates tend to run lower—they don't feel heat like men do. Researchers also note lots of guys wear suits and ties to work, while women might just wear light shirts or skirts. So it's no shock that plenty of women find the office too cold.

You might say, "Well, just have women wear more clothes!" But research shows this AC setup can waste up to 35% of energy, drag down worker efficiency due to an uncomfortable environment, spark "AC sickness" issues, and indirectly rack up avoidable cash losses.

So, beyond telling women to layer up, is there a better fix out there?

There's an old saying: "One kind of rice feeds a hundred kinds of people." Clearly, folks' habits and tastes come in all flavors. Even setting personal tastes aside, tons of factors still sway indoor thermal comfort, such as how many people are in a room, the number and type of computers or appliances running, or the kind of action going on—quiet work or heated meetings—all shift the room's temperature. Obviously, a fixed AC temp can't keep up with such wild use swings. What we need is a more flexible, adjustable, adaptive "smart" system.

With the rise and spread of technologies like the Internet of Things, cloud computing, and voice recognition, building control systems are becoming increasingly convenient and personalized. In residential buildings, this trait stands out especially clear. Using voice commands to run appliances, having smart systems recall our preferences or remotely steering the home vacuum robot from a phone while away—these scenes are now reality. In homes designed for seniors, monitoring gear, wearable health trackers, and comms devices can team up, sending help signals to caregivers fast if an older person falls, gets sick, or needs aid.

But in office buildings, we often share space with lots of people, so we can't control the whole environment as freely as at home. It's tough to have a smart system that precisely figures and manages each worker's personal micro-environment too. Instead of handing it all to computers, why not flip the script and give users the reins—put personalized control units at each desk so folks can tweak their own little zones?

Say you're sitting near a window—you could adjust curtains, blinds, reflectors, or lights to shift the light and heat around you. If you have a meeting, you could manage doors, windows, blinds, or sliding walls to dial in the sound around you. Weather changes—you could tweak the desk vent's airflow and temp to adjust your heat zone. You could also add air purification or dehumidifying to the airflow, boosting comfort even more. You've probably noticed we don't always need the whole space tuned the same—even in a tiny room, temperature and light vary corner to corner. Our surroundings aren't uniform—we just need to pick spots that suit the environment's quirks and tweak our own little patch. This is a kind of "smart" way to use buildings too.

Of course, the building's overall environment needs to stay in a decent range—not too extreme. Personal control builds on that decent base for finer, tailored tweaks. To achieve this effect, buildings need to be designed holistically with sound planning, on one hand, while individuals require a sufficient number of convenient personal control units, on the other. The Intelligent Workplace we talked about earlier is a prime example.

If every user tweaks and controls building equipment their own way, will it turn into a mess? Can this really save energy? Carnegie Mellon's CBPD actually ran an intriguing study on this.

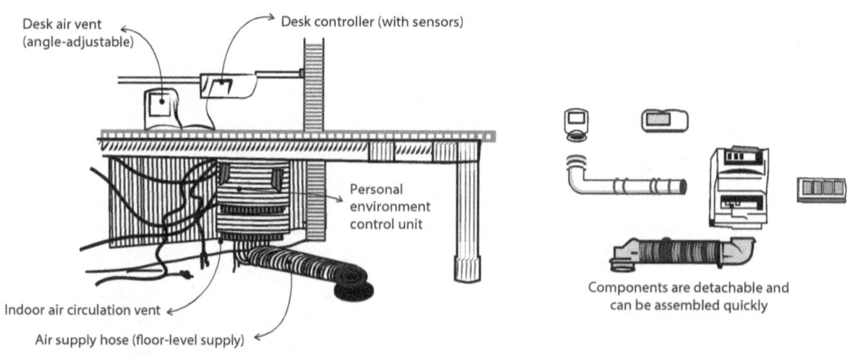

Personal environmental control unit used in the Intelligent Workplace

The research team picked an already-built, in-use office building as their case study, designing a smart interactive platform for its users. The platform's information is split into three levels.

Level 1: Information sharing—showing users building energy data so they can grasp it easily and check historical records from different time spans.

Level 2: Action tips—offering use suggestions based on how energy-hogging equipment (computers, lights, phones, fridges, water coolers, printers, coffee makers, etc.) runs in the building. For example, because computers guzzle energy, they suggest setting them to sleep after a bit of downtime

Level 3: Instant control—folding each energy-using device's on–off buttons right into the smart interactive system. Users can peek at energy data and tips and then hit a button below to flip things on or off right away.

During the preparation phase, the research team quietly logged the staff's daily appliance energy use, establishing a reference baseline. This phase lasted 14 weeks.

Then, the research team divided the participating staff into four groups—A, B, C, and D. Week 15, the test kicked off for real. Group A stayed the same as usual, with no changes made. Staff in Groups B, C, and D began using the smart interactive platform provided by the research team. This process lasted 13 weeks.

Among them, Group B could only access the first and second levels—viewing energy use of various building devices and getting some energy-saving tips. Results showed that under these conditions, the building's energy use dropped 13% below the baseline.

Group C could see the first, second, and third levels—adding on–off buttons for devices to Group B's setup, letting users easily click below tips to flip an appliance with one tap. This group reduced building energy use by 24% below the baseline.

Why did adding just an on–off button to the smart system cut energy use so much? The answer's obvious. Take this example: Users know a water cooler reheating eats power and should be shut off when idle, but it's a bit far, so they don't bother—or they're busy and forget later. There seems to be a gap between knowing what to do and actually doing it. But if they can one-tap the cooler off while seeing data and tips, users are likely to shut it down right then. Bridging the "knowing" and "doing" gap boosted results significantly right away.

What about Group D? Beyond seeing all the info and buttons from before, they got a calendar function to preset appliance on-off schedules for future days. Guess what? This time, the building saved 39% energy over the baseline! Weekend use dropped greatly since users could preset via the calendar to shut off equipment when they're not at the office.

And Group A, who never used the smart system—what changed? Results showed their energy use in the 13-week test phase still dipped 7% from the study's start. Why's that? Probably because, in the prep phase, Group A didn't know they were being studied, but during testing, just knowing someone was watching nudged them to perform better unconsciously. Aren't these subtle human thoughts and behaviors fascinating?

See, the research team didn't tweak the building's form, structure, or materials, nor did they swap in more energy-saving gear. Just showing users some info and handing them handy control buttons cut a building's energy use by 13% to 39%—pretty amazing, right? Turns out, fully respecting and trusting users, understanding their mindset and habits, and tapping their initiative can yield huge environmental gains.

Think it's over here? Not yet!

After the test phase wrapped, the team pulled the smart control system and kept tracking building energy use for 11 more weeks. Guess what—did folks' habits snap back overnight? Surprisingly, results showed energy use ticked up a bit, but for a while, it didn't bounce back to pre-study levels. This shows that during the time users had the smart control system, they shifted old habits, gradually building

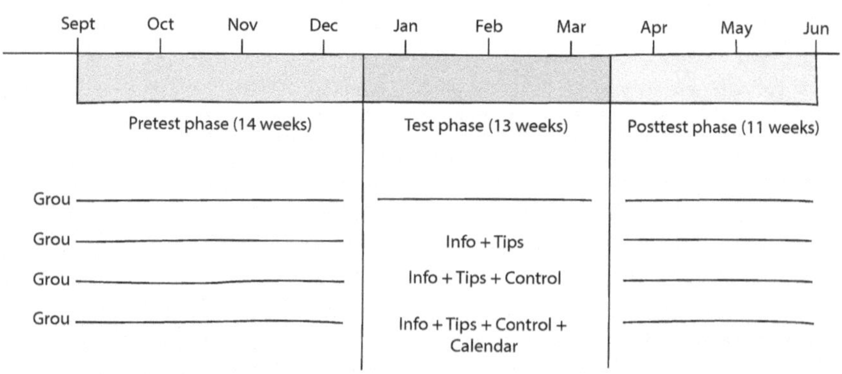

Experiment Setup

Experiment Results

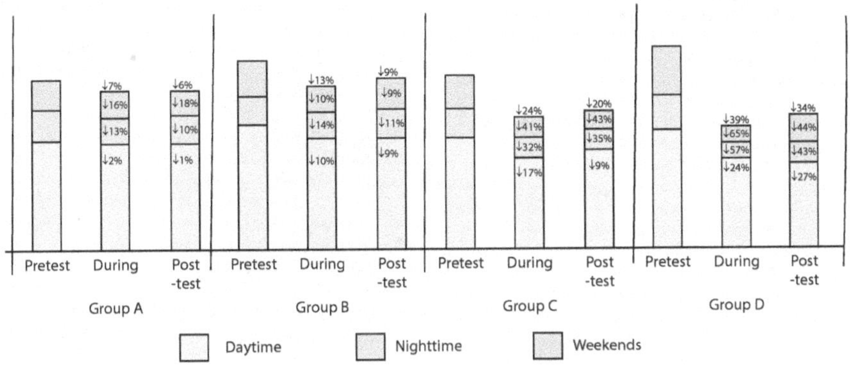

more eco-friendly behavior patterns. When users see their changes can slash building energy use with clear environmental perks, they're more eager to shift habits and build new, better ones. After all, everyone's got a drive to do better and chase bigger value and meaning—it just takes smart design and solid methods to fully spark that user initiative.

So has your confidence in your environmental impact grown a bit now?

You Know What?

Although human needs vary widely, they can be summed up with certain patterns. As early as 1943, the renowned American psychologist Abraham Harold Maslow proposed his hierarchy of needs theory, sorting human needs into five levels—"physiological needs," "safety needs," "love and belonging needs," "esteem needs," and "self-actualization needs"—known as the famous "Maslow's pyramid."

Clearly, while humans can be lazy, they're not always passive or idle. People crave good connections with their surroundings, yearn for respect and care, and, most crucially, long to realize their worth, prove their skills, and do what they find meaningful. Buildings serving users must also address these varied levels of need. Safety and comfort are just the base level—how can we make users feel a sense of belonging, fully respected, and capable of creativity and problem-solving? This requires technology and, even more, wisdom.

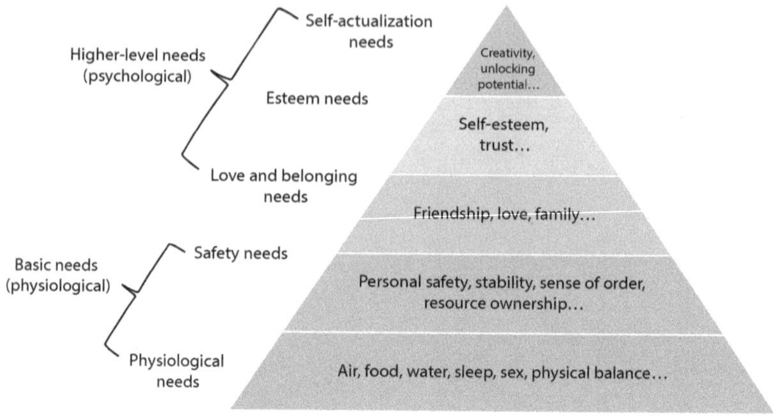

Maslow's pyramid of needs

07 Are Green Buildings Expensive?

We've covered tons of green building perks so far, but why does pushing them out seem so tricky? On this topic, your first thought is probably "Because green buildings cost a lot!" Solar panels—pricey! Wind turbines—pricey! Smart building systems—pricey! Eco-friendly materials—pricey! Are green buildings really just for the rich to enjoy? This chapter is where we dig into that.

07-1 Affordable Rural Green Buildings

Remember the raw-earth-building case from Chapter 1? In Yunnan's Ludian quake rebuild, villagers put up two types of homes—one's a simple brick-and-concrete mix, limited by cash and tech, with no insulation or outer decor; the other's a rammed-earth farmhouse using new tech, warm in winter, cool in summer, neat and comfy. Which of these two buildings do you reckon cost more?

Comparing these two, the latter, hands down, beats the former in quality, comfort, and looks. But on-site stats tell us, under market conditions back then, the brick-and-concrete home's construction cost[2] ran about 1,300 RMB per square meter, while the new rammed-earth farmhouse hit around 980 RMB per square meter.

Why does a pretty, sturdy, eco-friendly rammed-earth build cheaper per square meter than a rough, cold-in-winter, hot-in-summer brick mix? It's tied tightly to the building's setting and project context.

---

[2] Here, construction costs cover materials (including shipping) and labor for the foundation, walls, floors, roof, doors, windows, indoor wall leveling and whitewashing, plumbing, electrical, and insulation—not furniture, landscaping, design fees, or tech consulting costs.

Simple brick-and-concrete homes

New rammed-earth farmhouse

Guangming Village sits in the mountains, tough to reach. Outside industrial materials need a long haul to get to the village. Bricks and cement, cheap at first, cost a bunch more once shipped to the village. Plus, brick-and-concrete builds need a skilled crew or pro builders to put them up. Local villagers can't fully handle it—they've got to hire outside pros. After the quake, everyone wanted to rebuild fast—construction in the hit area spiked, driving up material prices, shipping costs, and crew rates due to short supply, but build quality didn't rise with the prices.

By contrast, the new rammed-earth farmhouse's main wall material—raw earth—came from quake rubble, recycled and refined with optimized mixes. The share of outside industrial materials dropped way down. Plus, since rammed earth is a local traditional method, the builders were village folks who could start work after basic training. This local-materials, local-tech, local-workers rebuild cut out relying on outside materials and crews and took less of a hit from market price swings. And because villagers were building for villagers—like teaming up to build for neighbors or kin—quality actually got a boost.

Just saving on build costs isn't the whole story. To break down green building economics, we've got to see where the money spent on building goes and which economy it lifts. In Guangming Village's brick-and-concrete homes, about 70% of the cash went to buying outside industrial materials, with only 30% or so on hiring workers—and a good chunk of those were from out of town. That means most of the villagers' building money flowed to outside markets, doing little for the local economy. But for the new rammed-earth farmhouse, over half the building cost went to hiring workers. It looks like labor costs beat out brick-and-concrete some, but because most of the materials and workers were local, building actually upped local incomes and jobs, doing more to boost the local economy and empower folks. In mountain rural areas with tough access, this "home-grown" economic model plays a big role in sustainable local growth.

The truth is, compared to a building's decades- or century-long life, construction costs are just a sliver of the total expense. Over years and years, operating costs stack up to a hefty sum. Like Guangming's brick-and-concrete builds with no insulation or energy-saving fixes, they're bound to need more fans, AC, heaters, and stoves for cooling and warmth. Rooms with poor daylight definitely need lights on longer to make up for it. But the new rammed-earth farmhouse, warm in winter and cool in summer, uses smart passive design to slash indoor heating and cooling time. Good natural light cuts down on artificial lighting too, naturally dropping operating costs big time.

Guangming Village post-quake rebuild cost comparison

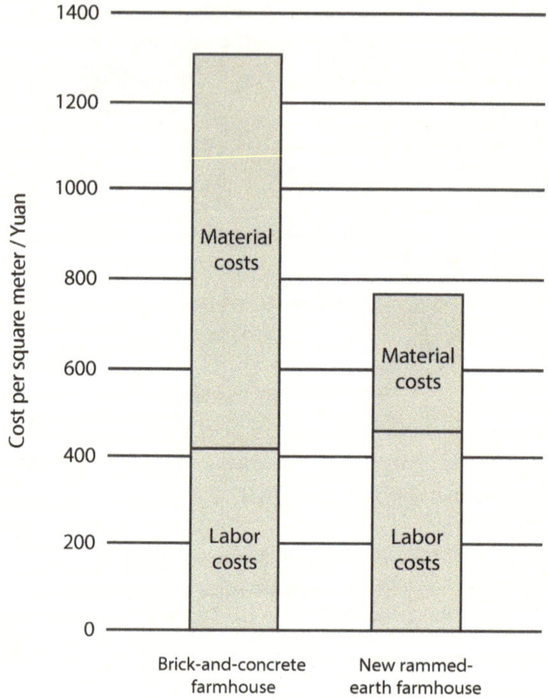

Then there are maintenance costs. Brick-and-concrete builds will need tile facings and renovations down the line. If another quake hits, causing warping or cracks, they'll need patching up. Local workers might not handle those jobs—outside help must be brought in, and materials come from afar too. But the new rammed-earth farmhouse's outer walls don't need extra decor—natural texture and beauty come built in. Thanks to innovative tech and design, it skips the usual roof and wall repair hassles. Even if fixes or expansions come up later, village builders can tackle most of it, naturally cutting upkeep costs a bunch. Even if fixes or expansions come up later, village builders can tackle most of it, naturally cutting upkeep costs a bunch.

Add it up, and the new rammed-earth farmhouse saves way more than just a few hundred bucks per square meter! For a remote, poor mountain spot like Guangming Village, building's got to save cash, up local skills, and fuel lasting local growth—that's the most cost-effective, sustainable green building.

You Know What?

Folks realize that in the usual industrialization and urbanization-driven growth model, economic gains always come with resource drain and environmental harm. Capital piling up spawns ever-bigger cities, while rural areas bear the brunt of resource use and environmental damage first. The hollowing out, marginalization, and decline of rural areas, sparking a slew of environmental, economic, and social issues, have drawn growing attention and reflection.

To tackle rural development woes, many developed nations started pumping outside cash and tech into villages through "exogenous development" to kick-start growth.

Rural roads, communications, and other infrastructure got a big upgrade, but folks noticed rural populations kept shrinking, and farm pollution and ecological decay didn't slow down. In remote, poor, rural spots with tough access and scarce resources, exogenous development's inputs and outputs often don't match up, making long-term sustainability a struggle.

By the late 1970s, the exogenous development model for rural areas caught a lot of flak. Exogenous development zeroed in on economic growth but overlooked rural core values. Picture this: If you're a rural kid, growing up seeing city demands shape rural setups, city tastes guide rural looks, and city tech blanket rural ways—even if your rural life improves, wouldn't you still dream of the big city? If they can't see their own unique strengths or local value, where's the drive for rural folks to protect the environment or build up their hometowns?

In the 1980s, folks began digging into and testing a new approach—"endogenous development"—more and more. As the name suggests, endogenous development doesn't lean hard on outside inputs or boosts—it trusts a region's natural, human, and cultural resources as the heart and key to growth. Endogenous development stresses that locals should lead growth, with the local economy and well-being powered by local natural resources, manpower, and regional culture. It puts more weight on local people's strength and role, boosting their ability to team up and join economic action, growing diverse, homegrown industries rooted in local history and culture—empowering locals, sparking their creativity, and driving sustainable environmental, social, and economic progress.

Of course, exogenous and endogenous development aren't oil and water. Each has perks, and they can even complement and lift each other up. Have you come across examples like this?

07-2 Green Buildings' "Slow Build, Big Payoff"

Are green buildings pricey or not? Let's crunch the numbers on green buildings now!

No denying it—some new tech, gear, and materials in green buildings, due to complex production or market rarity, usually cost more. If we take a regular building's build cost as the "baseline cost," then the extra cost to build a green one is its "incremental cost."

Incremental costs break down into incremental construction costs, design consulting costs, and maintenance management costs. Incremental design consulting costs mean the cash for hiring pro–green building designers or consultants—compared to construction costs, this slice is pretty small. Market analysis shows green building maintenance costs don't differ much from regular ones—the incremental bit is barely noticeable—so the big talker's incremental construction costs.

A 2012 study dug into 55 Chinese building projects that passed green building reviews and got green labels—30 residential, 25 public. Findings showed green residential projects' incremental construction costs ranged from 0.43 to 168.9 RMB per square meter, while green public projects hit 5.72 to 306.07 RMB per square meter. Compared to older data, both incremental costs show a clear downward trend year by year. This proves China's green buildings have grown fast over the past few years, with design skills, market balance, tech picks, and cost control all maturing day by day.

Researchers also found high-performance gear like energy-saving lights, ACs, and solar water heaters are super common now—stable in cost and quality and not tricky to install or maintain. The incremental cost for using this gear has become very low—some are near zero. But for trickier, less common materials like solar panels and wind turbines, incremental costs stay pretty high. Still, as time rolls on, tech improves, and products spread; more high-performance gear and renewable systems will hit regular homes; and incremental costs will drop too.

Worth noting—these buildings' incremental costs don't always line up straight with their green building rating stars. That means in green building ratings, higher-star buildings don't always mean higher incremental costs. This points to a key issue: green building efficiency isn't directly tied to cost. Sometimes, with smart, clever overall design, you can hit ideal results even without high incremental costs.

After covering incremental costs, you're probably thinking there's "incremental benefits" too, right? Exactly—green buildings pack plenty of those over their full life cycle. For example:

- Compared to regular buildings, savings on energy and water costs during operation
- Financial incentives owners or developers might snag (tax breaks, subsidies, bonuses, etc.)
- Economic gains from boosted worker productivity in green buildings
- Corporate image and brand value built by constructing or using green buildings
- Macroeconomic benefits from green buildings (spurring industry growth, adding jobs, etc.)
- Profits from selling surplus renewable energy or carbon credits

Among these incremental benefits, the easiest to grasp and most noticed is the first one. Take a typical Beijing office building as an example—if a 10,000-square-meter building uses 40–60 kWh per square meter yearly (1 kWh = 1 kilowatt-hour) and hits a 50% energy-saving goal after upgrades, it'd save 164,000 to 246,900 RMB in electric bills each year. If that building has 20 years left, it could save over 4.5 million RMB total. Pretty impressive, right?

If we weigh incremental costs against incremental benefits, we'll see green buildings aren't pricey at all. There are plenty of research papers on this—specifics vary, but roughly, China's green building costs run 5%–7% above regular ones now, and those costs pay back in 5–8 years as the benefits roll in. And for the rest of the building's life, those extra benefits keep piling up.

Of course, we're still facing a bunch of challenges along the way. For instance, a building's investors, operators, and users are usually different folks. How do investors footing the incremental costs directly reap the benefits? That takes careful business model planning and clever setup. Plus, making incremental costs and benefits more transparent, calculable, and visible is a big factor in whether decision-makers go green. And how do we get city power grids to link up easier with green buildings' renewable systems, handle and buy back that energy better? How do we push carbon trading so more emission cuts get noticed and paid off? These are directions we all need to work on together. I'm confident that as times evolve and society progresses, policies and markets will lean more toward green economies—folks will naturally pick healthier, comfier lifestyles, and green building economic perks will shine even brighter.

08 Grading Green Buildings

Earlier, we noted some projects that passed green building reviews and earned green building labels. Ever wonder what kind of review that is? Do green buildings need to pass a test and get a certificate too? This chapter is where we'll chat about how to grade green buildings.

08-1 Green Buildings Need a Certificate Too

In green building's growth, folks realized these projects are complex system jobs, needing architects, owners, contractors, and other stakeholders to team up. In this multiparty, multilevel collaboration, everyone's goals and judging standards might differ. When making decisions, one side thinks this way, another that—if there's a clear review and certification system to gauge building performance, it's easier to get everyone on the same page and push the project forward better. Plus, a review system helps all sorts grasp green building perks, aiding promotion and winning market and policy backing. That's why tons of countries and regions started building green building review systems.

The world's first green building review system popped up in 1990 in the United Kingdom—the "Building Research Establishment's Environmental Assessment Method" (BREEAM). It's run by the U.K.'s Building Research Establishment (BRE), a century-old hub for building environment science. As the "granddaddy" of building environment reviews, it's been a key base and reference for tons of later green building systems worldwide.

In 1998, the U.S. Green Building Council rolled out a market-driven review system—"Leadership in Energy & Environmental Design" (LEED). From the get-go, it baked in market vibes, pushing market-driven action for green building equipment, materials, and related industries. Consider this: if you buy and use LEED-approved green materials, you score bonus points in the LEED review. This setup not only nudges architects to pick top-notch, efficient new materials but also spurs the market to crank out more of them, kicking green building markets into a healthy loop. Plus, LEED built a solid training and certification system, letting keen folks learn LEED through courses and snag pro credentials. Certified pros can then jump into projects better, driving the review and certification work. This design and operation scored LEED huge wins in market reach and recognition, making it the world's most widely used green building review system.

In 2002, Japan dropped its own system—"Comprehensive Assessment System for Building Environment Efficiency" (CASBEE). It brought fresh twists to scoring and result displays, inspiring tons of later systems to borrow from it. (How'd it innovate? We'll get to that next section—little teaser for now.)

In the 1990s, the green building idea started picking up steam in China. China rolled out a series of guides and systems like the "Key Points and Technical Guidelines for Green Ecological Residential Communities" and the "Green Olympic Building Assessment System." Drawing on studies of big-name global systems, plus regional, targeted pilots and wrap-ups, China launched its national green building standard—"Green Building Evaluation Standard"—in 2006, setting up green building labels from there.

According to the World Green Building Council, there are about 60 different green building assessment systems worldwide. From 2016 to 2021, the global floor area certified by various green building systems jumped from 104 million square meters to 420 million square meters. Quite a few building projects have won certifications from multiple different systems at once. Plus, these data only cover systems run by green

China's 2021 updated green building labels

building councils in various countries and regions—toss in ones set up and managed by other groups, and the count is way higher. Clearly, green building assessment systems play a huge role in pushing green buildings and sustainable development forward. With green building systems, markets, and policies all steering the way, green buildings are in a booming phase.

08-2 Assessment Systems Aren't Simple

Usually, green building assessment systems boil down to three main parts:

First, a tightly knit set of standards describing environmental performance. (That's the system's metric framework.)

Second, assigning points to different metrics—hit a specific performance standard, and you score. (That's the scoring method.)

Third, showing the final score and the building's environmental performance through a specific approach. (That's how results are presented.)

The first part is like the system's "skeleton." A system's framework sets its scope and the logic behind its standards. That scope defines the building types, scales, and timeframes the system applies to. Early systems often only worked for certain building types, like public or

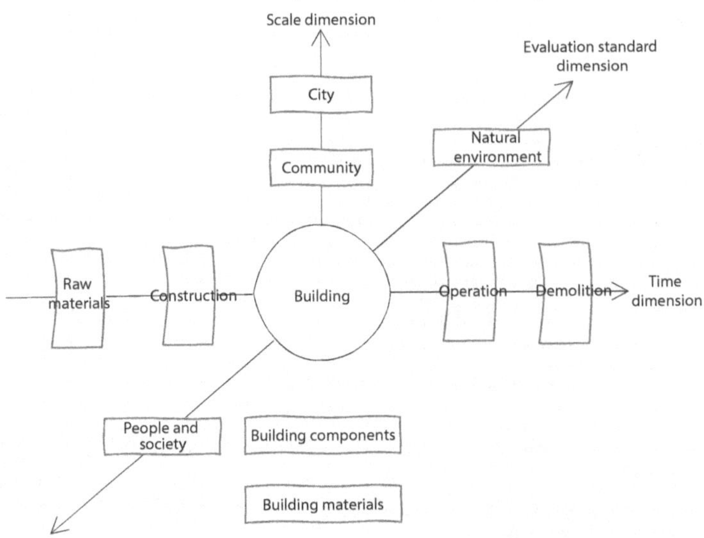

Three dimensions of building environment assessment

Table 8.1. Assessment scopes of some green building systems

BREEAM (UK)	• New construction • Renovation and refurbishment • Existing buildings • Communities
LEED (USA)	• Building design and construction • Interior design and construction • Building operation and maintenance • Neighborhood development • Residential • Cities and communities
CASBEE (Japan)	• Residential scale • Building scale • Urban district scale • City scale
China Green Building Evaluation Standard	• Evaluation targets are single buildings or building complexes • Evaluation happens after building completion, with pre-evaluation possible at the design stage

residential ones. Time-wise, they might only fit new builds, not retrofits of existing ones. Scale-wise, they'd just handle single buildings, not smaller indoor designs or bigger blocks or city plans. As time rolled on and systems evolved, most got polished into full sets of standards, aiming to cover different phases, types, and scales of building environments.

In assessment standards and logic, folks first focused on pollution control, energy saving, resource conservation, and boosting indoor environment quality—metrics leaned hard on energy use, water resources, and indoor air quality. As sustainability ideas sharpened and spread, people caught on that environmental, social, and economic sustainability are a must-have trio that play off each other—so lots of systems started adding social and economic sustainability metrics. China's Green Building Evaluation Standard, in its 2019 update, swapped the old "land saving," "energy saving," "water saving," "material saving," and "environmental protection" framework—aka "four savings, one eco"—for five performance pillars: "safety and durability," "health and comfort," "life convenience," "resource saving," and "environmental livability," putting more weight on social sustainability and spotlighting a "people-first" approach.

Table 8.2. Indicator system of the 2019 Updated Green Building Evaluation Standard

Safety and durability	I. Safety II. Durability	8 control items, 9 scoring items
Health and comfort	I. Indoor air quality II. Water quality III. Acoustic and light environment IV. Indoor thermal and humidity environment	9 control items, 11 scoring items
Life convenience	I. Travel and accessibility II. Service facilities III. Smart operation IV. Property management	6 control items, 13 scoring items
Resource saving	I. Land saving and land use II. Energy saving and energy use III. Water saving and water resource use IV. Material saving and green building materials	10 control items, 18 scoring items
Environmental livability	I. Site ecology and landscape II. Outdoor physical environment	7 control items, 9 scoring items
Improvement and innovation		10 bonus items

In recent years, many countries and regions have developed green building assessment systems tailored to their natural environments and social contexts. Several nations have also adapted established systems like BREEAM or LEED, modifying them to suit their own green building evaluations. Existing assessment systems are progressively evolving, refining frameworks and metrics to enhance scientific rigor, practicality, and alignment with sustainable development objectives.

In terms of scoring and calculating results, current assessment systems primarily fall into two types: The first aggregates scores from individual evaluation items (multiplied by weights if applicable), summing them to yield a total score. This resembles the scoring method of an examination paper. Well-known systems such as BREEAM and LEED employ this calculation approach.

The second type is the Q-L assessment method used by Japan's CASBEE system. Q represents building environmental quality and performance, encompassing indoor environmental quality, building service quality, and outdoor environmental quality, while L denotes building environmental loadings, including energy consumption, resource and material use, and environmental impact. CASBEE defines the ratio of quality to load as building environment efficiency (BEE). If a project scores lower on environmental load (L) and higher on quality (Q), its BEE rating increases. Based on this methodology, CASBEE provides highly intuitive result charts, enabling readers to quickly grasp a project's characteristics, strengths, and weaknesses while facilitating comparisons across multiple projects. This scoring approach has been adopted and integrated into many subsequent assessment systems.

When it comes to naming assessment results, different systems vary quite a bit. China's green building evaluation standard splits results into four levels—meeting all control items hits the basic level. From there, one-star, two-star, and three-star ratings mark low to high, with three stars as the top grade. LEED's assessment results come in four levels: "Certified," "Silver," "Gold," and "Platinum." BREEAM's results are split into six grades: "Unclassified," "Pass," "Good," "Very Good," "Excellent," and "Outstanding."

Most systems on the market set a baseline—above it, you pass and get graded; below it, you're just marked as failing across the board. For practical use, this setup is plenty handy. But some systems even grade the failures, with results like "Poor" or "Very Poor." Maybe this better captures a building's performance, making comparisons or redesigns easier, although it sounds a tad strict. Different systems' baselines often tie to their country or region's typical conditions, not some uniform standard. Results from different systems can't just be compared side by side. For example, is LEED's Silver better than BREEAM's Excellent? Tough to say off the bat.

So, test scores matter, sure, but we've got to keep a cool, rational head about them. What's your mindset when tackling all kinds of tests?

08-3 Beyond the Score Is Where It Starts

You might find this a bit dry by now; why drone on about assessment systems? Truth is, I just want you to see that judging a green building's success involves tons of factors and all sorts of standards. Different systems come with their own backstories, focuses, and scopes. When using these tools, we should first get a grip on their pros and cons, not just chase scores blindly.

Let me tell you another story. While chasing my PhD, I once tried using LEED—this widely loved system—to assess a rural post-disaster rebuild we'd done. I figured our project, with its solid take on environmental, social, and economic sustainability for local growth, ought to score pretty well. But I was bummed to find out that 75% of LEED's metrics didn't even apply to our rural project. Some metrics were built for urban scale, density, and infrastructure, useless for rural settings;

LEED v4 BD+C: New Construction and Major Renovation

Project scorecard

Project name:

Date:

满足	?	不满足			
			得分点	整合过程	1
0	0	0			16
			得分点	LEED 社区开发选址	14
			得分点	敏感土地保护	1
			得分点	高优先场址	3
			得分点	周边密度和多样化土地使用	5
			得分点	优质公共交通连接	5
			得分点	自行车设施	1
			得分点	停车面积减量	1
			得分点	绿色机动车	1
0	0	0			10
			先决条件	施工污染防治	必要项
			得分点	场址评估	1
			得分点	场址开发－保护和恢复栖息地	3
			得分点	空地	1
			得分点	雨水管理	3
			得分点	降低热岛效应	2
			得分点	降低光污染	1
0	0	0			11
			先决条件	室外用水减量	必要项
			先决条件	室内用水减量	必要项
			先决条件	建筑整体用水计量	必要项
			得分点	室外用水减量	2
			得分点	室内用水减量	4
			得分点	冷却塔用水	2
			得分点	用水计量	1
0	0	0			33
			先决条件	基本调试和查证	必要项
			先决条件	最低能源表现	必要项
			先决条件	建筑整体能源计量	必要项
			先决条件	基础冷媒管理	必要项
			得分点	增强调试	1
			得分点	能源效率优化	13
			得分点	高级能源计量	1
			得分点	能源需求反应	2
			得分点	可再生能源生产	3
			得分点	增强冷媒管理	1
			得分点	绿色电力和碳补偿	2
0	0	0			13
			先决条件	可回收物存储和收集	必要项
			先决条件	营建和拆建废弃物管理计划	必要项
			得分点	减少建筑生命周期中的影响	5
			得分点	建筑产品分析公示和优化－产品环境要素声明	2
			得分点	建筑产品分析公示和优化－原材料的来源和采购	2
			得分点	建筑产品分析公示和优化－材料成分	3
			得分点	营建和拆建废弃物管理	3
0	0	0			16
			先决条件	最低室内空气质量表现	必要项
			先决条件	环境烟控	必要项
			得分点	增强室内空气质量策略	2
			得分点	低逸散材料	3
			得分点	施工期室内空气质量管理计划	1
			得分点	室内空气质量评估	3
			得分点	热舒适	1
			得分点	室内照明	2
			得分点	自然采光	3
			得分点	优质视野	1
			得分点	声环境表现	1
0	0	0			6
			得分点	创新	1
			得分点	LEED Accredited Professional	1
0	0	0			4
			得分点	地域优先：具体得分点	1
			得分点	地域优先：具体得分点	1
			得分点	地域优先：具体得分点	1
			得分点	地域优先：具体得分点	1
0	0	0			110

认证级：40至49分，银级：50至59分，金级：60至79分，铂金级：80至110分

LEED scoring table example

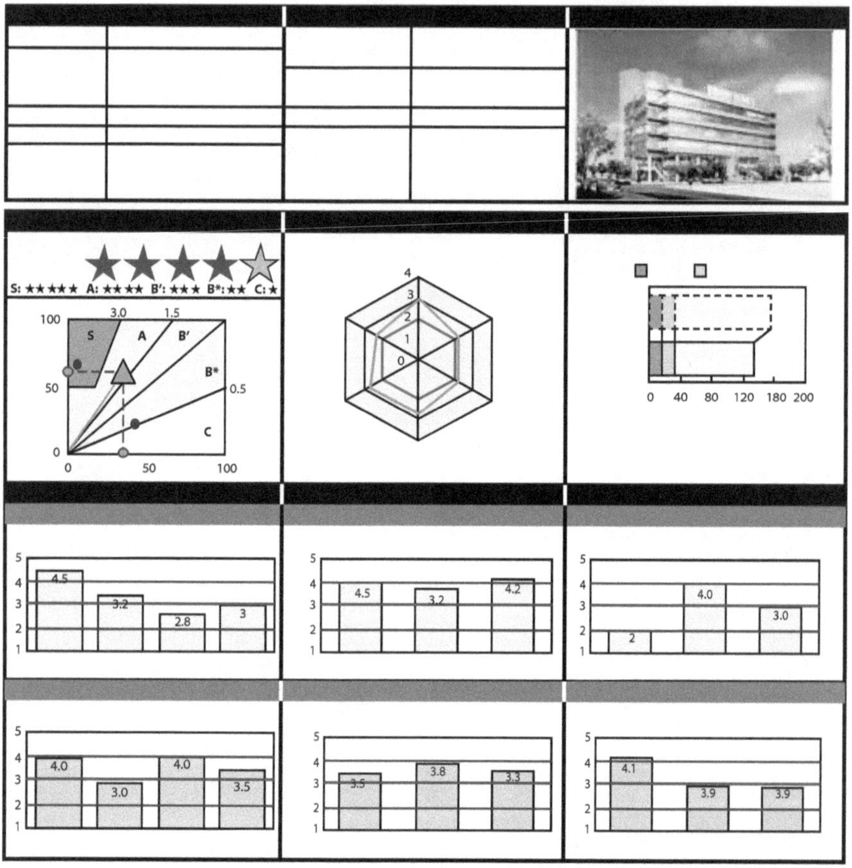

CASBEE assessment result example

some green materials lists skipped nonstandard local materials like rammed earth and bamboo; some indoor quality standards fit city buildings and lifestyles, clashing with rural ways; and some needed computer simulations or field measurements that just weren't doable in rural projects. That "faceplant" got me rethinking green building systems big time, opening a whole new view. I didn't doubt our project's quality just because it tanked in the system—instead, my PhD thesis reexamined rural green building and sustainability's backdrop and ideas, pitching a framework I thought better fit rural needs. Although still in the research phase, that framework guided our team's rural studies and projects for over a decade after. Through constant theory–practice loops and polish, we all became surer of our path.

From saving energy and guarding the environment to sustainability, green building has spanned decades. Folks' takes on green buildings and what they expect keep shifting—new ideas and practices pop up all the time. Sometimes, we can even set our own goals and standards based on a project's context and traits, going beyond current systems to nudge it closer to our ideal. Assessment systems are tools to help us hit sustainability, but they're definitely not the finish line.

On this journey of researching and practicing green building, I always hear voices saying, "What's the point? Can you really change the world or save humanity?" True, one person's effort makes tiny ripples—if we tackle every little task aiming to shift the world or save folks, failure and helplessness might haunt us continually. Here's how I see it: Picking different lifestyles is a way we live in this world—a state that first makes us feel safe, comfy, and healthy, then gives us self-respect, belonging, the joy of loving and being loved, and the thrill of self-fulfillment. Remember Maslow's pyramid? In a sustainable life-style, all our need levels get fuller play, so bit by bit, we lean toward it without even noticing. Our own physical and mental health, happiness, and well-being are really two sides of the same coin as human sustainability. To some degree, we'll all find a bit of common ground on this. So what thoughts will you carry into picking your future lifestyle?

References

[1] Hou Jiyao, Wang Jun. *Chinese Cave Dwellings* [M]. Zhengzhou: Henan Science and Technology Press, 1999.

[2] Shi Kehui, Hu Xuesong. *Yunnan Vernacular Architectural Culture* [M]. Nanjing: Southeast University Press, 2003.

[3] Mu Jun, Zhou Tiegang, Wang Shuai, et al. *Guide to New Rammed Earth Green Housing Construction Techniques* [M]. Beijing: China Architecture & Building Press, 2014.

[4] Ye Zuda, Li Hongjun, Song Ling. *Cost-Benefit Analysis of Green Building Technology and Economics in China* [M]. Beijing: China Architecture & Building Press, 2013.

[5] Wan Li, Wu Enrong, Chi Xin'an. From Post-Disaster Reconstruction to Rural Revitalization: The "One Expert, One Village" Guangming Village Demonstration Project [J]. *Architectural Techniques*, 2017 (263): 24–29.

[6] Li Hongtu. Environmental Pollution and Control During the British Industrial Revolution [J]. *Exploration and Debate*, 2009 (2): 60–64.

[7] Zhou Yuehui. Three Stages of Western Urbanization [J]. *Theoretical Guide*, 2013 (2): 42–42.

[8] Xue Ming. Creating Green Living: The BedZED Eco-Community [J]. *Journal of Chongqing Jianzhu University*, 2004 (z1): 61–66.

[9] Hong Yingjin. A Model Green Enterprise: O'right Green Factory [J]. *Architectural Journal (Taiwan)*, 2014 (9 Special Issue): 72–77.

[10] Zhou Guanwu, Jiang Yuyan. A New Architectural Path for the German Reichstag [J]. *Creative Design Source*, 2011 (2): 30–35.

[11] Fang Jin. Endogenous Rural Development Model: Practical Exploration, Core Features, and Reflective Expansion [J]. *Journal of China Agricultural University (Social Sciences Edition)*, 2018 (1): 24–34.

[12] Zhang Wenming, Zhang Zhimin. Resources, Participation, Identity: Logic and Path Choices for Endogenous Development in Rural Revitalization [J]. *Social Sciences*, 2018 (11): 75–85.

[13] Zhang Huanzhou, Huang Chaochao, Zhou Yongguang. A Review of Endogenous Development Models [J]. *Journal of Zhejiang University (Humanities and Social Sciences Edition)*, 2007 (2): 61–68.
[14] Wan Yimeng, Xu Rong, Huang Tao. Comparative Analysis of China's Green Building Evaluation Standard and U.S. LEED [J]. *Building Science*, 2009 (8): 6–8.
[15] Li Ju, Sun Daming. *Statistical Analysis of Incremental Costs of Green Buildings in China* [C]. Beijing: 5th International Conference on Intelligent, Green Building, and Building Energy Conservation, 2009: 66–71.
[16] Huang Xianming. *Research on Ecological Economic Optimization Issues of Green Buildings* [D]. Beijing: Tsinghua University, 2006.
[17] Ministry of Ecology and Environment of the People's Republic of China. *Concepts and Significance of Biodiversity* [EB/OL], 2010–01–14 [Accessed 2022–09–12].
[18] O'Right International. *Green Architecture: The Beauty of Harmony Between Buildings and Nature* [EB/OL], 2019–10–09 [Accessed 2022–09–12].
[19] *Construction Industry Zero Carbon Building* [EB/OL]. Hong Kong: Construction Industry Council, 2020–04–13 [Accessed 2022–09–12].
[20] SPARK Architects. *Clarke Quay Redevelopment Project* [EB/OL], 2016–06–07 [Accessed 2022–09–12].
[21] O'right International. *2016 O'right Corporate Social Responsibility Report* [R/OL], 2017–12–01 [Accessed 2022–09–12].
[22] Xu Junmin. *Research and Application of Green Roofs and Vertical Greening in Hong Kong* [R/OL]. Hong Kong Institute of Surveyors Young Surveyors Group Technical Seminar, 2014–05–19 [Accessed 2022–09–12].
[23] Wang Qingqin. *Introduction to China's Green Building Standards* [R/OL]. Beijing: China Academy of Building Research Co., Ltd., 2019–07–18 [Accessed 2022–09–12].
[24] Zhang Yongsheng, Lu Meihui, Cui Muyang. *Post-Disaster Seismic Survey of Houses* [N/OL]. Beijing: Beijing News, 2014–08–11 [Accessed 2022–09–12].
[25] Bai Yu, Su Hexian, Lai Zhengcong. *Research Report on Seismic Simulation Shaking Table Tests of New Rammed Earth Wall Houses* [R]. Kunming: Yunnan Institute of Earthquake Engineering, 2015.
[26] Ministry of Housing and Urban-Rural Development. *"13th Five-Year Plan" for Building Energy Conservation and Green Building Development* [R], 2017.
[27] Ministry of Housing and Urban-Rural Development. *Guide to Energy-Saving Retrofits for Existing Residential Buildings* [R], 2012.
[28] *CASBEE Assessment Results for Chiba Port Tower (Japanese Company)* [OL]. Japan Enterprise Investment Co., Ltd., 2010–01–22 [Accessed 2023–07–03].
[29] Carson R. *Silent Spring* [M]. Boston: Houghton Mifflin Company, 1962.
[30] World Commission on Environment and Development. *Our Common Future* [M]. Oxford: Oxford University Press, 1987.

[31] Vale B, Vale R. *The Autonomous House: Design and Planning for Self-Sufficiency* [M]. London: Thames and Hudson Ltd., 1976.
[32] Givoni B. *Man, Climate and Architecture* [M]. London: Applied Science Publishers, 1976.
[33] Deyan S. *Norman Foster: A Life in Architecture* [M]. New York: The Overlook Press, 2010.
[34] Malcolm Q. *The Norman Foster Studio: Consistency Through Diversity* [M]. London and New York: E & FN Spon, an Imprint of Routledge, 1999.
[35] Busenkell M, Schmal PC. *WOHA: Breathing Architecture* [M]. London: Prestel Publishing, 2012.
[36] Bingham-Hall P. *WOHA: Selected Projects Volume 2* [M]. Singapore: Pesaro Publishing, 2015.
[37] Clements-Croome DJ. *Creating the Productive Workplace* [M]. London: E & FN Spon, 2000.
[38] Kamal MA. An Overview of Passive Cooling Techniques in Buildings: Design Concepts and Architectural Interventions [J]. *Civil Engineering & Architecture*, 2021 (55): 84–97.
[39] Chance T. Towards Sustainable Residential Communities; the Beddington Zero Energy Development (BedZED) and Beyond [J]. *Environment & Urbanization*, 2009 (21): 527–544.
[40] Kingma B, Lichtenbelt WVM. Energy Consumption in Buildings and Female Thermal Demand [J]. *Nature Climate Change*, 2015 (5): 1054–1056.
[41] Fisk WJ, Rosenfeld AH. Estimates of Improved Productivity and Health from Better Indoor Environments [J]. *Indoor Air*, 1997 (3): 158–172.
[42] Maslow AH. A Theory of Human Motivation [J]. *Psychological Review*, 1943 (50): 370–396.
[43] Hartkopf V, Loftness V, Mahdavi A, et al. An Integrated Approach to Design and Engineering of Intelligent Buildings—The Intelligent Workplace at Carnegie Mellon University [J]. *Automation in Construction*, 1997 (6): 401–415.
[44] Cole RJ. Building Environmental Assessment Methods: Clarifying intentions [J]. *Building Research and Information*, 1999 (27): 230–246.
[45] Sanner B, Kabus F, Seibt P, et al. Underground Thermal Energy Storage for the German Parliament in Berlin. *System Concept and Operational Experiences* [C]. The World Geothermal Congress, Antalya, Turkey, April 2005.
[46] Chi X, Bai W, Ng E. *Study on the Rammed-Earth Building in the Post-Earthquake Reconstruction of Southwest China: A Case Study in Ludian* [C]. Terra 2016 YJJ World Congress on Earthen Architecture, Lyon, France, June 2016.
[47] Abergel T, Dean B, Dulac J. *Global Status Report 2017* [R]. United Nations Environment Programme, 2017.

[48] Hodge J. *BedZED Seven Years on* [R]. BioRegional, 2009.
[49] Lazarus N. *BedZED: Toolkit Part II: A Practical Guide to Producing Affordable* [R]. BioRegional, 2003.
[50] Department of Architecture, The Chinese University of Hong Kong. *Feasibility Study for Establishment of Air Ventilation Assessment System Final Report* [R]. Planning Department, 2005.
[51] Dorgan Associates. *Productivity and Indoor Environmental Quality Study* [R]. National Energy Management Institute, 1993.
[52] Gibberd J. *Integrating Sustainable Development into Briefing and Design Processes of Buildings in Developing Countries: An Assessment Tool* [D]. Pretoria: University of Pretoria, 2003.
[53] Heringer A. *Vision* [EB/OL], 2020–01–01 [2022–09–12].
[54] Hockerton Housing Project [EB/OL], 2021–03–16 [2022–09–12].
[55] Department of Architecture. *The Commons* [EB/OL]. ArchDaily, 2016–12–18 [2022–09–12].
[56] HHP. *Case Study: Retrofit for the Future* [EB/OL], 2012–07–01 [2022–09–12].
[57] McGee C. *Shading* [EB/OL]. Canberra: Australian Government, 2013–01–01 [2022–09–12].
[58] Bird M. *An Architect's Sustainable Dreams* [EB/OL]. London: China Dialogue, 2006–10–30 [2022–09–12].
[59] Anderson J. *Embodied Carbon & EPDs* [EB/OL]. GreenSpec, 2011–01–01 [2022–09–12].
[60] United Nations. *What Is Climate Change?* [EB/OL], 2021–04 [2022–09–12].
[61] Earth Science Communications Team at NASA's Jet Propulsion Laboratory. *Global Climate Change* [EB/OL], 2022–09–01 [2022–09–12].
[62] The Center for Building Performance and Diagnostics (CBPD) [EB/OL]. Pittsburgh: Carnegie Mellon University School of Architecture 2008 [2022–09–12].
[63] BREEAM [EB/OL], 2022–09–01 [2022–09–12]. Watford: BER Group.
[64] LEED Rating System [EB/OL]. Washington, DC: U.S. Green Building Council 2000–01–01 [2022–09–12].
[65] Comprehensive Assessment System for Built Environment Efficiency (CASBEE) [EB/OL]. Japan Sustainable Building Consortium and Institute for Building Environment and Energy Conservation, 2015–12–02 [2022–09–12].
[66] World Green Building Council Annual Report 2020/21 [R/OL]. London: World Green Building Council, 2021 [2022–09–12].
[67] Belluck P. *Chilly at Work? Office Formula Was Devised for Men* [N/OL]. New York: The New York Times, 2015–08–3 [2022–09–12].
[68] Ward L. Energy Dashboards Enter the Office Cubicle Studies Give Workers Tools to Monitor and Reduce Their Individual Energy Use [N/OL]. *The Wall Street Journal*, 2013–09–22 [2022–09–12].

Afterword

Whether you skimmed it like a dragonfly or savored every bit, thanks for finishing this book. If you've made it to the afterword, I guess you don't think I'm too shabby as a guide.

I'll let you in on something without worrying you'll laugh—when my editor first pitched this book idea, I was pregnant with my second kid. During this book's writing, my baby was born, and my editor became a mom too. Along the way, I often felt downright ashamed of my snail-like pace. But with perfectionism nagging at me, I'd often freeze up, scared I wasn't good enough, and just stall out. Writing this book was basically my epic battle with procrastination.

So here, I must give special recognition to my editor, Ms. Liu Yilin, for her endless patience and support—her gentle grit kept me from throwing in the towel. Thanks go to my PhD mentor, Professor Wu Enrong at the Chinese University of Hong Kong, and my like-minded research crew—senior and junior peers, teammates, and work buddies. From a volunteer in rural projects, to Professor Ng's student, to teaming up with him postgrad on rural sustainability research—I've learned not just the "how" of research from him, but the "why" of life. The whole team gave me room and backing to tackle concerns that might seem pointless but felt deeply meaningful to me.

Big thanks to Prof. Brenda Vale, Prof. George Baird, Prof. Ray Cole, Prof. Adrian Pitts, Prof. Lutz Katzschner, Prof. Lamkhee Poh, and Prof. Ye Zuda, who traveled far to teach the MSc in Sustainable and Environmental Design at the Chinese University of Hong Kong. A ton of the knowledge and inspiration in this book springs from your brilliant teaching and living examples. You're my role models in academic work and on life's path.

Thanks to my parents, husband, and two kids for putting up with me always taking the road less traveled. Raising kids has given me heaps of fresh insights into buildings and life.

Finally, huge thanks to my instrument teachers—Teacher Niu Tianyuan, Teacher Cheng Long, Mr. Shaqiu TTY—and my fellow learners; when I was wrestling with my procrastination or worn out from juggling work and life roles, your company and music brought me comfort and strength.

"Sustainability" is both a concept and a way of living. I hope this book's brought you some fun and food for thought, and that you'll carry this wisdom into your future, living as a sustainable, ever-renewing, fully vibrant person.

Wan Li
Midsummer 2023

FSC
www.fsc.org
MIX
Papier | Fördert
gute Waldnutzung
FSC® C083411

Kolibri 360 GmbH
Ferdinand-Jühlke-Straße 7
99095 Erfurt, Deutschland
produktsicherheit@kolibri360.de